For my good friend
Jay!

*Jeff Nelson*

# Die Kaiserlichen Schutztruppen und Polizeitruppen für Afrika

## Imperial German Colonial Troops & Police in Africa

Uniforms · Equipment · Weapons · Organization

by Reinhard Schneider

translated by Chris Dale

Pedestrian Press

Toledo, Ohio

All rights reserved
© 2005, 2023 by Reinhard Schneider

No part of this book may be reproduced
or transmitted in any part or by any means, electronic
or mechanical, including photocopying, recording, or by any
information storage and retrieval system without permission
in writing from the publisher.

This book was translated by Chris Dale
from the original German edition, entitled
*"Die Kaiserliche Schutz- und Polizeitruppe für Afrika
– Felduniform, Ausrüstung, Bewaffnung"*,
published in 2005 by
Druffel & Volwinckel-Verlag,
Stegen am Ammersee.  ISBN 3-8061-1162-6

For additional information on this subject,
visit www.GermanColonialUniforms.co.uk

ISBN 978-0-9626506-1-1
Printed in Germany.
Book design by Ingo Priebe/Graphikdesign and
Jeff Nelson/Pedestrian Press.  Art direction & cover design
by Jeff/P. Press.  Front cover illustration by Anton Hoffman, from
"Der Deutsche Soldat mit Waffe und Werkzeug – Armee Bilderbuch",
Cl. Attenkofer'sche Verlagsbuchhandlung, Straubing, ca. 1904.

First Printing

Pedestrian Press
Toledo, Ohio

www.PedestrianPress.com

## Author's Note

The author's interest in the German colonies and the Schutztruppen was awakened in 1980, when he met his wife, Yvonne Roscher-Schneider. She is the granddaughter of Dr. Max Roscher, seen above. He was the last Post and Telegraph Inspector of Togo, and on 2 August 1914 he was serving as *Leutnant der Reserve* and head of military intelligence. During the First World War he participated in the attempt to repel the invading British forces, and, that effort having failed, was instrumental in the destruction of the radio station at Kamina.

The author has tried to present the Imperial German Colonial Forces as accurately as possible. Combining varying segments of information into one cohesive volume was difficult, and certainly this work makes no claim to completeness. However, the author hopes that overall he has managed to give a detailed picture of the soldiers and police of the German colonies in Africa.

## Translator's Note

Regarding the titles of most unit types and ranks, I have for ease retained German terminology throughout most of this book. The German Colonial Troops were called the *Schutztruppen* or Protectorate Troops. German Colonial Paramilitary Police Forces were called the *Polizeitruppen*. Both of these German titles will already be familiar to many readers and are more memorable than their lengthier English translations. In the case of ranks, I have also retained German titles. This is because translated rank titles often cause confusion. For example, while a German *Leutnant* and a British Lieutenant might appear the same, they are actually not equivalent ranks.

I would also like to thank Dan Baune for his patient help and advice on several of the more complex sections of the book.

## Publisher's Note

This is a purely historical study, and in no way an endorsement of colonialism. History is unfortunately rife with man's inhumanity to man, and the exploitation of native Africans and their lands by European powers in the 19th and 20th centuries was certainly no exception.

## Sources and Acknowledgements

Information on uniforms was largely taken from the reminiscences of former *Schutztruppe* officers, some of which contained passages on the organisation and uniforms of the *Schutztruppen*. Regulations from contemporary orders and communications were also consulted. Images within this book come primarily from the author and Jeff Nelson. We were also fortunate to obtain a magnificent set of 8 photo albums assembled by *Hauptmann* Ernst Nigmann during his service in German East Africa, many images from which are reproduced here for the first time.

The author would like to thank Jeff Nelson for his friendship, and for his determination to expand and ultimately publish this English edition. In turn, Jeff and Reinhard want to thank Chris Dale and Ingo Priebe for their intelligent input and the invaluable skills they brought to the project. And of course Jeff in Toledo thanks his three compatriots in Berlin and London for their patience with his perfectionism and paper-centric ways these past five years.

A debt of gratitude is owed to many for their help and support: Ulrich Ender, Harry Fakner, Jason Farrell, Dieter and Marion Gerlach, Siegenot Haluschka, Tim Healy, Martin Hedler, Andreas M. Schultze Ising, Jan Kröger, Jan Kube, Patrizia Maiotti, Jan Müller, Margaret and Ted Nelson, Arne Schoefert, Google Translate, Christian Unsinn, Helmut Weitze and Berliner Zinnfiguren.

# Contents

**Establishment of the German Colonies**

**The Role of the Schutztruppe**

**Part I – German South-West Africa**
(*Deutsch Südwestafrika*)

- 9   Commanders of the *Schutztruppe*
- 9   Formation of the *Schutztruppe*
- 11   Uniforms of the *Schutztruppe*
- 43   Equipment of the *Schutztruppe*
- 48   Weapons of the *Schutztruppe*
- 53   Organisation of the *Schutztruppe* in 1914
- 54   Formation of the *Landespolizei*
- 56   Uniforms of the *Landespolizei*
- 59   Equipment of the *Landespolizei*
- 60   Weapons of the *Landespolizei*
- 61   African Police Auxiliaries (*Polizeidiener*)
- 64   Organisation of the *Landespolizei* in 1914
- 64   The First World War in German South-West Africa
- 100   *Schutztruppe* Anthem

**Part II – German East Africa**
(*Deutsch Ostafrika*)

- 105   Commanders of the *Schutztruppe*
- 105   Formation of the *Schutztruppe*
- 122   Uniforms of the *Schutztruppe*
- 128   Equipment of the *Schutztruppe*
- 128   Weapons of the *Schutztruppe*
- 129   Uniforms of the African Troops (*Askari*)
- 144   Equipment of the *Askari*
- 146   Weapons of the *Askari*
- 147   Marching on Campaign (*Kriegsmärsche*)
- 148   Training of the *Askari*
- 152   Organisation of the *Schutztruppe* in 1914
- 153   Uniforms, Equipment and Weapons of the *Polizeitruppe*
- 153   Organisation of the *Polizeitruppe* in 1914
- 163   The First World War in German East Africa
- 189   *Heia, Safari* Anthem

**Part III – Cameroon**
(*Kamerun*)

- 193   Commanders of the *Schutztruppe*
- 193   Formation of the *Schutztruppe* and *Polizeitruppe*
- 196   Uniforms, Equipment and Weapons of the *Schutztruppe*
- 207   Organisation of the *Schutztruppe* in 1914
- 208   Cameroon *Polizeitruppe*
- 211   The First World War in Cameroon
- 218   Map of *Schutztruppe* Garrisons in Cameroon
- 219   Cameroon *Schutztruppe* March

**Part IV – Togo**

- 223   Commissioners and Governors of Togo
- 223   Formation of the *Polizeitruppe*
- 225   Uniforms, Equipment and Weapons of the *Polizeitruppe*
- 228   Governor's Decree concerning *Polizeitruppe*, Local Police, Border Guards, Chieftain Policemen
- 240   Organisation of the *Polizeitruppe* in 1914
- 240   The First World War in Togo
- 241   Togo Poetry Verses

**Part V – Medals, Campaign Clasps, and Recognitions of Service**

- 245   South-West Africa Medal (*Südwestafrika-Denkmünze*)
- 247   South-West Africa Medal Campaign Clasps
- 248   Colonial Medal (*Kolonialdenkmünze*)
- 249   Colonial Medal Campaign Clasps
- 250   Combatant's Merit Medal (*Kriegerverdienstmedaille*)
- 252   Service Commendations and Commemorative Items
- 274   Imperial German Colonial Flag
- 275   Map of German Protectorates in Africa

**Glossary** and **Selected Bibliography**

*Imperial German Colonial Troops & Police in Africa*

28 August 1914, Special Edition of *Breslauer General-Anzeiger* newspaper.
The headline reads 'The Conflicts in our Colonies'.

# Establishment of the German Colonies

By the middle of the nineteenth century, the great powers of Europe had divided almost the entirety of Africa among themselves. Thus, England, Belgium, France, Portugal, Spain, Italy, and the Netherlands all had large overseas colonial possessions.

The German states had meanwhile been embroiled in a series of Wars of Unification in the mid-19th century and had found no time for acquisitions outside the continent of Europe. After the Prussian King Wilhelm unified Germany in 1871, twelve years passed before Prussian Chancellor Otto von Bismarck turned his attention to securing colonies for Germany.

In 1883, the Bremen merchant Adolf Lüderitz acquired lands around Angra Pequena in south-west Africa. To prevent England from taking over the territories, on 24 April 1884, Bismarck placed them under the protection of the German *Reich*.

Around the same time, German merchants on the west African coast in Togo and Cameroon requested Imperial support, as they often felt threatened by tribal chiefs who had been incited by the British.

Dr. Gustav Nachtigal hoisted the German flag in Bagida on 5 July 1884 and then in Lome the next day. These actions claimed Togo as a German Protected State. After negotiations with the tribal princes, Nachtigal also raised the German flag in Cameroon on 14 July 1884.

In early November 1884, the armoured frigate SMS Elisabeth arrived in the Pacific Ocean off the island of New Britain, where she was met by the gunboat SMS Hyäne (Hyena). In the next few days the warships claimed the island groups of New Britain (later New Pomerania), New Ireland (later New Mecklenburg), Lavongai (later New Hanover) and the Duke of York Islands (later New Lauenburg) as protectorates of the German Empire. The area was then named the Bismarck Archipelago.

The German Trade and Plantation Company and the New Guinea Company were founded between 1882 and 1884, and were the driving forces for the occupation of the northern part of New Guinea, which had not yet been claimed by a European state. Dr. Otto Finsch was chosen for the task. On 3 November 1884, in the presence of Dr. Finsch, the commanders of the SMS Elisabeth and SMS Hyäne raised the German flag on Maputi, a small island in the Bismarck Archipelago, and placed the country under the protection of the German Empire.

The region was named Kaiser-Wilhelmsland and on 17 May 1885, Kaiser Wilhelm I granted the New Guinea Company a letter of protection, effectively giving it sovereignty over the new colony.

Meanwhile, South-Eastern New Guinea was placed under British protection. The Dutch had settled the western part of the island at the beginning of the 19th century. With final frontier adjustments made by treaty in 1886, New Guinea was divided among the three European powers.

While the German Empire was establishing itself in the South Seas, Dr. Carl Peters,

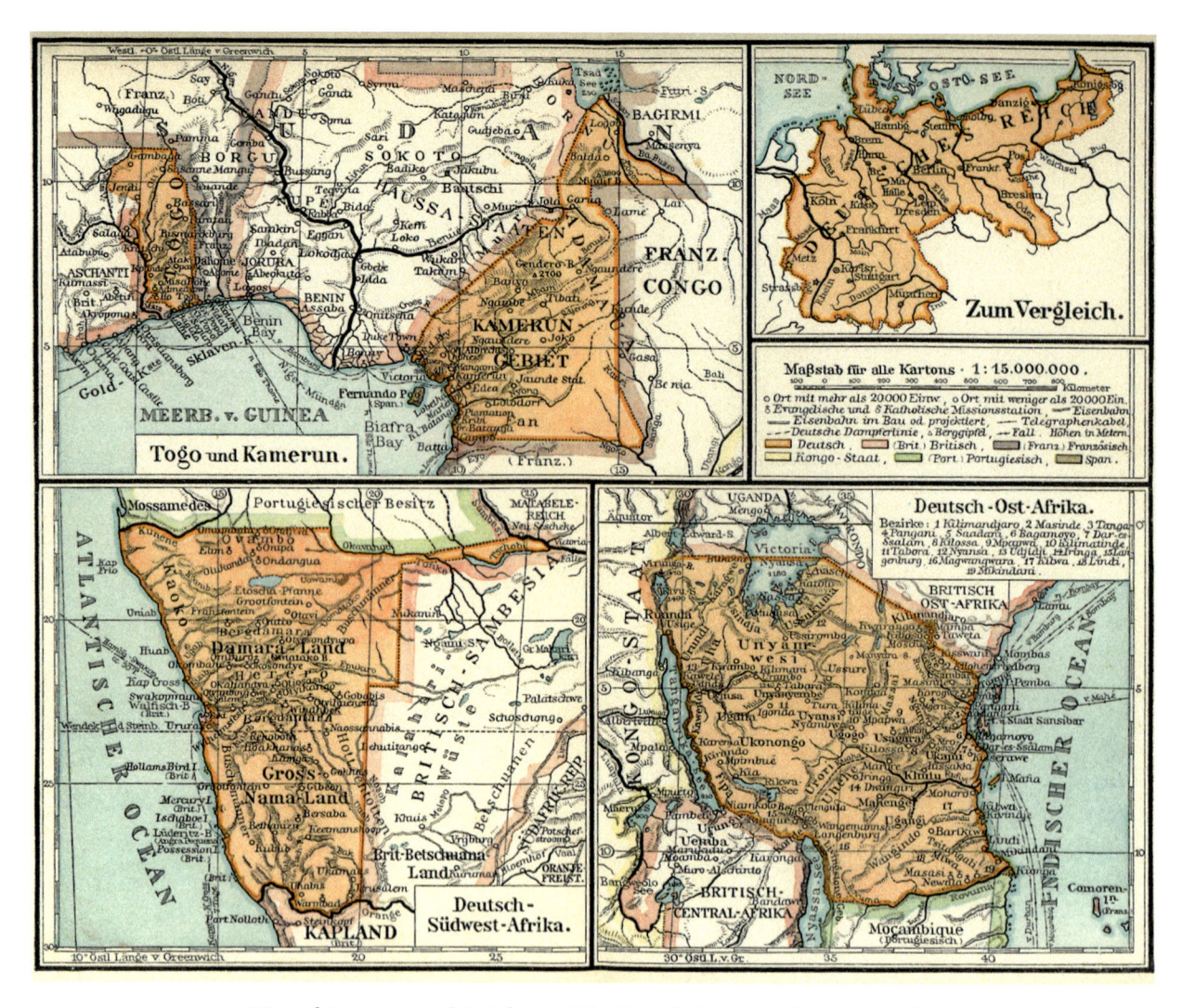

Map of Germany and its African Colonies, all shown at the same scale.

Dr. Karl Ludwig Jühlke and Count Pfeil claimed areas of Eastern Africa in present-day Tanzania on behalf of the Society for German Colonization. On 10 November and 17 December 1884, the territories of Ussagara and Ukami (an area roughly the size of Ireland and Wales, or slightly smaller than the State of Pennsylvania) were acquired by treaties. Kaiser Wilhelm I signed a letter of protection for this area on 27 February 1885.

The following example illustrates how cheap it was to acquire land at that time. On 8 April 1885, the brothers Clemens and Gustav Dehnhardt concluded a protection and land purchase contract with the Sultan of Witu, Ahmed ibn Fumo Bakari (known as Simba, the Lion) in modern Kenya. For the cost of 50 rifles, some cloth, glass beads and 1,000 Maria Theresa Thaler, 25 square kilometres of land was bought and a German protectorate claimed over the Sultanate. At the time Wituland was approximately 3,000 square kilometres, two-thirds the size of the German *Reich*.

On 1 July 1890, Germany and Great Britain signed the Zanzibar Treaty, under which Germany took possession of Helgoland Island in the North Sea and added the Caprivi Strip to German South-West Africa. In return, Germany made concessions concerning its colonial possessions adjacent to Angola and Rhodesia (present-day Zambia). Contrary to popular belief, it was not the island of Zanzibar that was the subject of a barter transaction, but Wituland, although Germany did promise to refrain from pursuing her interests in Zanzibar.

Incidentally, during the colonial period, textbooks listed the highest mountain in the German Empire as Mount Kilimanjaro.

For some time Germany had sought an ice-free harbour on the Chinese coast as a coaling base for the Imperial Navy's South Sea Squadron. In response to the murder of two German missionaries in China in 1897, German naval troops occupied Kiautschou Bay, the strategic significance of which had already been pointed out by the geographer Ferdinand von Richthofen. On 6 March 1898, German and Chinese authorities concluded a 99-year lease on the territory of Kiautschou Bay, with a harbour to be built at Tsingtao. This was similar to a deal struck that same year in which Britain acquired Hong Kong for 99 years. The province of Shantung is still known to us today for the old German brewery whence Tsingtao beer is sold all over the world.

The final addition to the German Empire was made in 1899, with the annexation of Western Samoa.

At the time of the European Powers' taking possession of and dividing territories amongst themselves, very few thought to take the indigenous populations into account. Borders were often drawn arbitrarily on a map, rather than being based on existing tribal or ethnic groups. Several examples of this could be seen on maps of the German colonies:

1) In Togo, much of the border with French Dahomey was a straight line.

2) In German East Africa, the border with British East Africa was formed by two straight lines on either side of German-held Mount Kilimanjaro.

3) In South-West Africa, the aforementioned Caprivi Strip jutted out 450 kilometres to the East. Modern Namibia still retains this geographical oddity within its borders.

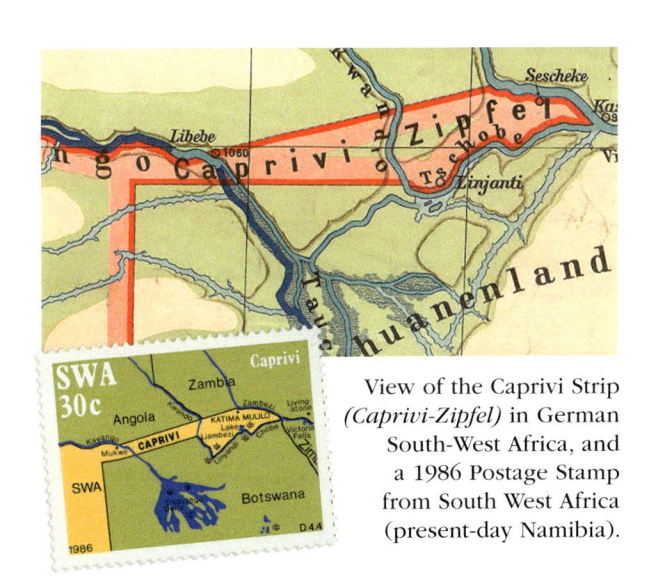

View of the Caprivi Strip *(Caprivi-Zipfel)* in German South-West Africa, and a 1986 Postage Stamp from South West Africa (present-day Namibia).

## The Role of the *Schutztruppe* in the German Colonies in Africa

The first *Schutztruppe* regulation of 25 July 1898 stated that "The *Schutztruppe* are to serve to maintain public order and security in the African colonies and in particular to combat the slave trade." The main task of the *Schutztruppe* was therefore to protect against indigenous uprisings, but also included occasional police duties and assistance in the cultural development of the country, for example in land surveying.

There were many local uprisings against German colonial rule in Africa, the most notable of which were the Herero and Nama Rebellions in German South-West Africa (1904 – 09) and the near-simultaneous Maji-Maji Rebellion in German East Africa. The crushing of these uprisings by the *Schutztruppe* resulted in the deaths of tens of thousands of the African population.

A war against the armies of the other European colonial powers was neither envisioned nor prepared for in terms of strength, armament, or the stockpiling of weapons and ammunition for the *Schutztruppe*. Instead, the German Empire relied on the maintenance of the Congo Treaty of 1885, which guaranteed the neutrality of the African colonies in the event of a future European War.

Proposals for Colonial Coats of Arms from Secretary of State Dr. Wilhelm Solf. Executed by Heraldic Painter Max Block in 1913, and approved by Kaiser Wilhelm II in July 1914, two weeks before the outbreak of the First World War.

Colonial commanders and governors frequently advocated for reinforcement and reorganisation of their security forces, pointing out that overseas forces might be called upon to fight against a European power. Their efforts, however, were in vain.

The doctrinaire principle that "the fate of the colonies will be decided in the North Sea" did not contemplate a role for the *Schutztruppe* in a European war. Thus, the militarization of the colonies was not addressed. Borders were left unfortified, which meant no fortifications existed to defend the colonies or act as springboards to invade neighbouring territories. Nor were any naval bases created in Africa for the overseas fleet.

On 2 August 1914, the First World War broke out with a series of declarations of war. Great Britain's declaration of war on Germany was transmitted on 4/5 August via Lome in Togo to Lüderitzbucht in German South-West Africa.

Trusting in the neutrality of the African colonies guaranteed by the Congo Treaty, the Colonial Secretary of State in Berlin, Dr. Wilhelm Solf, sent a telegram to German East Africa on 2 August 1914 to calm nervous settlers, stating that the colonies would be excluded from the danger of war. This trust proved to be misplaced.

Top row, left to right: Colonial Coats of Arms for German South-West Africa and German East Africa. Bottom row, left to right: Coats of Arms for the Colonies of Cameroon and Togo.

Photos: Siegenot Haluschka

Mossamedes
Portugiesischer Besitz
Kunene
Ovambo
Ondjiva
Elim
Onipa
Olukonda
Ondangua
Okavango
Kap Frio
Kaoko
Etoscha-Pfanne
Grootfontein
Uniab
Franzfontein
Otavi
Buschmänner
Bergdamara
Damara-Land
Huab
Omaruru
Omatako B.
Okombahe
Okosondye
Epukiro
Herero
Kap Gross
Okahandya
Otyosasu
Swakopmund
Otyimbingwe
Gobabis
Walfisch-B
(Brit.)
Windhoek
Bergdamara
Naossanabis
Rehoboth
Hoakhanas
Rehutitang
Hollams Bird I.
(Brit.)
Gross-
Gibeon
Nama-Land
Mercury I.
(Brit.)
Ischaboe I.
(Brit.)
Bersaba
Bethanien
Keetmanshoop
Lüderitz-B
(Angra Pequena)
Possession I.
(Brit.)
Kubub
Hottentotten
Brit-Betsc
Lan
Ukamas
Uhabis
Warmbad
Jerusalem
Orange
Port Nolloth
Steinkopf
KAPLAND
(Brit.)
ATLANTISCHER OCEAN
Östl Länge v. Greenwich
20

# Part I
# German South-West Africa
*(Deutsch Südwestafrika)*

Map of German South-West Africa, 1906.

## Commanders of the *Schutztruppe*

1894 – 1895  *Major* Curt von Francois

1897 – 1904  *Oberst* Theodor Leutwein

1904 – 906  *Generalleutnant* Lothar von Trotha

1906 – 1907  *Generalmajor* Berthold von Deimling

1907 – 1911  *Oberst* Ludwig von Estorff

1912 – 1914  *Oberstleutnant* Joachim von Heydebreck

1914 – 1915  *Oberstleutnant* Victor Franke

## Formation of the *Schutztruppe*

The South-West African *Schutztruppe* underwent many changes over time. It began in 1888 as an armed protection unit for the German Colonial Society (*Deutsche Kolonialgesellschaft für Südwestafrika*), and consisted of two former German army officers, five former army NCOs, and 20 native Baster and Herero. This scant troop, stationed at Otjimbingwe on the Swakop River about 100 kilometres west of Windhoek, was intended as the nucleus of a larger indigenous force to be created at some future point. Their first task was the protection of gold deposits that had been discovered on the territory of the German Colonial Society in 1887. These gold finds did not prove worthy of mining and the unit, which was deemed unsuitable for expansion due to its military weakness, was disbanded in October 1888.

*Major* Leutwein, Commander of the *Schutztruppe* for German South-West Africa at the outbreak of the Herero Rebellion in 1904.

In 1889, *Hauptmann* Curt von Francois (at the time surveying the border in Togo), together with his brother, *Leutnant* Hugo von Francois of the 26th Infantry Regiment, received orders to create a *Schutztruppe* for South-West Africa. Initially it was intended to be a mercenary unit in the style of the East African *Wissmanntruppe*, a private force under state control and discipline. This newly formed unit consisted of 21 active-duty men from the German army and thirteen from the reserve. All were volunteers and signed a private contract obliging them for three years' service.

The *Francois-Truppe* uniform was made to the order of Mission Director Büttner and

L to R: Seals of the Overall *Schutztruppe* Command
and German South-West African *Schutztruppe* Command.

consisted of a yellow-brown, velvet-like ribbed Manchester corduroy tunic and trousers, long brown boots with spurs and a brown belt with cartridge pouches. The hat was made of grey felt with a wide brim, the right side of which was flipped upwards and held in place by a German cockade. A greatcoat, woollen blanket, bread bag and water bottle completed the men's equipment. The riding equipment consisted of the horse's bit, bridle, halter, saddle and felt saddle cloth.

While the garments proved themselves and remained basically the same, the initial armament of a Mauser carbine, revolver and short dagger did not.

The carbine was replaced by the *Gewehr M71/84* infantry rifle, of which *Hauptmann* von Francois brought sufficient numbers from Togo. In 1890 these were replaced by the *Gewehr M88* rifle, which was the followed by the *Gewehr M98* at the time of the Herero Rebellion in 1904. This rifle became the armament of the entire South-West African *Schutztruppe*. The dagger was soon replaced by the infantry bayonet.

By an Order of His Majesty in Council (*Allerhöchste Kabinettsorder* or *AKO*) decree of 16 September 1911, it was declared retroactively that the official formation date of the South-West African *Schutztruppe* was 16 April 1889. The *Schutztruppe* was founded during the period of commercial administration of the colony, which transitioned into a military administration in 1891. Likewise, the unit was initially a private force, of 30 to 50 volunteers who committed themselves to the person of the commander.

This was increased to 350 men in 1893, but further reinforcements were soon needed, and it became clear that this original arrangement was no longer sustainable. So the original *Truppe des Reichkommissars v. Francois* became the *Kaiserliche Schutztruppe* and thus (along with the Imperial Navy) the only force under Imperial command.

In 1894, it was organised into four field companies with fifteen officers and 500 men, to which a field artillery battery, a skilled workers unit (*Handwerkerabteilung*) and a railway control (*Kommando der Eisenbahntruppe*)

*Oberst* Deimling, *Schutztruppe* Commander, 1906 – 1907.

## Uniforms of the *Schutztruppe*

The **Cloth Uniform** (*Tuchuniform*) or **Home Uniform** (*Heimatuniform*) was introduced on 11 March 1897 for the *Schuztruppen* of all colonies to wear when in Germany. It consisted of a tunic based on that of the Prussian infantry with Guard braid (*Litzen*), but in pale grey with Swedish-style cuffs and colony-coloured collar and piping. In the case of German South-West Africa, this was cornflower blue. The white metal buttons of the tunic bore the Imperial crown. Shoulder straps took the form of four-fold mohair cord in black/white/red sewn together for NCOs and other ranks. These were again fastened with a small Imperial crown button. NCO rank insignia was as for the Prussian army. The trousers were in matching grey, piped also in the colony colour of cornflower blue.

The **Corduroy Tunic** (*Kordwaffenrock*) resembled the pale-grey Home Tunic in most respects but was made of sand grey corduroy. All insignia was the same as on the Home Tunic. The front was fastened by eight white metal buttons with the Imperial crown. After washing and being bleached by the sun, the corduroy became paler and less uniform in color.

The **Field Tunic** (*Feldrock*) had a jacket-like shape and was made of yellow brown khaki drill. There was a 12 cm split down the rear skirt. The stand-and-fall collar and buttonless Swedish cuffs were made of the same material and had cornflower blue piping, which also edged the front flap of the tunic. At the collar was an eyelet with a closing hook.

was added in 1896. At that time it was still subordinate to the Imperial Navy Office, but after 1896 command was transferred to the Colonial Office and the *Kommando der Kaiserlichen Schutztruppen*. All ranks were supplemented by volunteers from the army contingents of the German states. At times of special need, reservists residing in the colony could be required to serve. Depending on the situation in the colony, the strength of the *Schutztruppe* increased in coming years to 700 and then to 900 men, for an area one-and-a-half times the size of Germany.

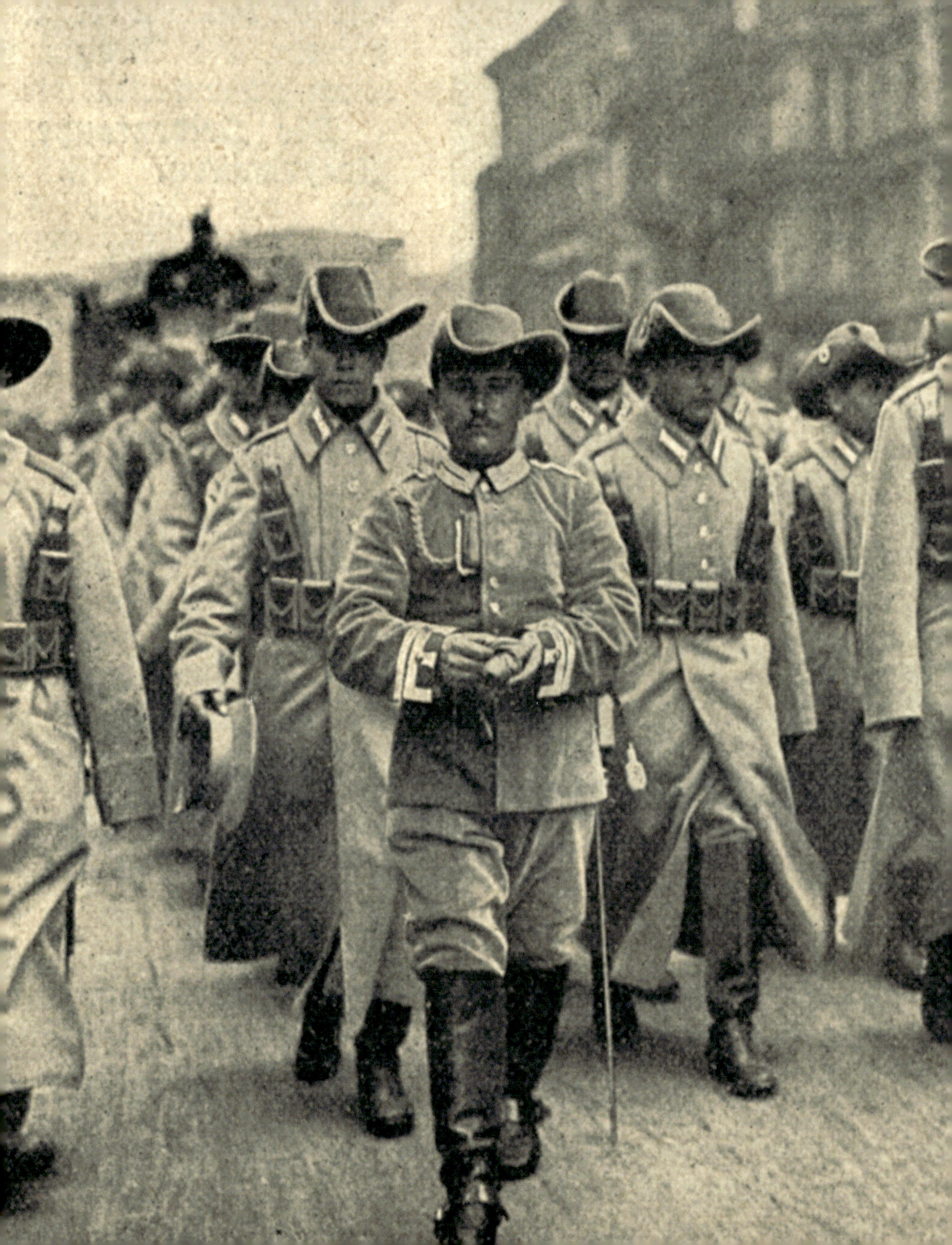

*Schutztruppe* Soldiers in Berlin, soon to leave for German South-West Africa. 24 March, 1904.

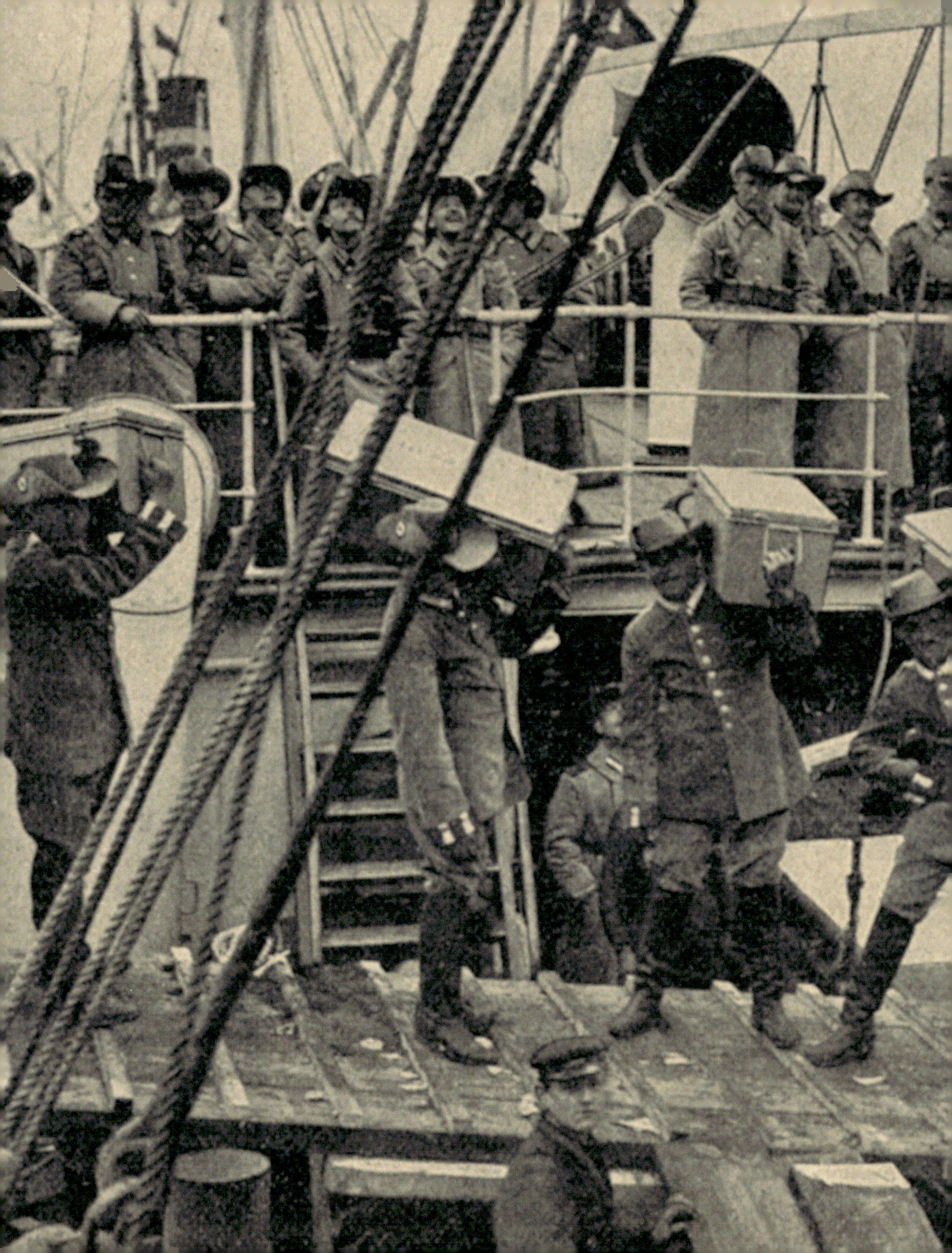

*Schutztruppe* Soldiers on board the Steamship *Lucie Wörma*n en route to German South-West Africa, 1904.

# Die Deutsche Schutztruppe

From a lithograph by **Moritz Ruhl, Leipzig**.

These figures represent uniforms authorized 11. June 1894, including corduroy and khaki drill uniforms. Among uniform details later discontinued are Kepi-style cap and pointed Polish-style cuffs. These illustrations show corduroy tunics in shade of grey as stated in uniform regulations, however surviving examples show them to have been more brown than grey. Figures have been moved to improve fit on page and German captions removed.

**Medical Officer** wearing the Corduroy Uniform

**Corporal** wearing the Greatcoat

**Hospital Assistant** wearing the Corduroy Uniform

**Hospital Assistant** wearing the Khaki Uniform

für Südwest-Afrika. 1894

| **Corporal** | **Veterinarian** | **Junior Paymaster** | **Sergeant Major** |
|---|---|---|---|
| wearing the Corduroy Uniform as Parade Dress | wearing the Corduroy Uniform | wearing the Corduroy Uniform | wearing the Khaki Uniform |

Die Deutsche Schutztruppe

**Officer**

wearing the
Greatcoat

**Officer**

wearing the Corduroy
Uniform as Parade Dress

**Officer**

wearing the
Khaki Field Uniform

**Officer**

wearing the Corduroy
Uniform as Field Dress

## für Südwest-Afrika. 1894

These figures represent uniforms authorized 11 June 1894, including corduroy and khaki drill uniforms. Among uniform details later discontinued are Kepi-style cap and pointed Polish-style cuffs. These illustrations show corduroy tunics in shade of grey as stated in uniform regulations, however surviving examples show them to have been more brown than grey. Figures have been moved to improve fit on page and German captions removed.

From a lithograph by **Moritz Ruhl, Leipzig.**

**Soldier**

wearing the Corduroy Uniform as Field Dress

**Bugler**

wearing the Corduroy Uniform as Field Dress

**Corporal**

wearing the Khaki Field Uniform

**Schutztruppe für**

From a lithograph by **Moritz Ruhl, Leipzig.**

These figures represent uniforms authorized 19. November 1896, including corduroy and khaki drill uniforms. These illustrations show corduroy tunics in shade of grey as stated in uniform regulations, however surviving examples show them to have been more brown than grey. They also show cuff buttons on the khaki tunic, but this is incorrect. Some figures were moved to improve fit on page, and German captions have been removed.

| **Medical Officer** | **Hospital Assistant** | **Hospital Assistant** | **Hospital Assistant** |
|---|---|---|---|
| wearing the Khaki Uniform | wearing the Greatcoat | wearing the Corduroy Uniform | wearing the Khaki Uniform |

# Südwestafrika. 1896

| **Corporal** | **Veterinarian** | **Paymaster** | **Sergeant Major** |
|---|---|---|---|
| wearing the Corduroy Uniform as Parade Dress | wearing the Corduroy Uniform | wearing the Khaki Uniform | wearing the Khaki Uniform |

Schutztruppe für

| **Officer** | **Officer** | **Officer** | **Officer** |
|---|---|---|---|
| wearing the Greatcoat | wearing the Corduroy Uniform as Parade Dress | wearing the Khaki Uniform | wearing the Corduroy Uniform as Field Dress |

# Südwestafrika.     1896

These figures represent uniforms authorized 19. November 1896, including corduroy and khaki drill uniforms. These illustrations show corduroy tunics in shade of grey as stated in uniform regulations, however surviving examples show them to have been more brown than grey. They also show cuff buttons on the khaki tunic, but this is incorrect. Some figures were moved to improve fit on page, and German captions have been removed.

From a lithograph by **Moritz Ruhl, Leipzig.**

**Soldier**

wearing the
Corduroy Uniform

**Bugler**

wearing the
Corduroy Uniform

**Junior Gunsmith**

wearing the
Corduroy Uniform

Offizier.   Unteroffizier.
Die Uniformierung und Ausrüstung der deutschen Schutztruppe für das südwestafrikanische Schutzgebiet.

Uniforms and Equipment of the *Francois-Truppe*, by H. Lüders. *Illustrierte Zeitung*, 9 November 1889.

The tunic was fastened by six white metal buttons bearing the Imperial crown (on a grained background for officers, smooth for NCOs and other ranks). Both sides of the tunic had an external breast pocket, which sloped slightly towards the middle, roughly level with the second and third buttons. The pockets had square-cut flaps and a vertical centre fold. There were two pockets (without centre fold) at the bottom of the skirt, which had straight flaps. These pocket flaps were closed with small buttons with the Imperial crown. The *Feldrock* had the same shoulder straps as worn on the *Tuchuniform*.

*Schutztruppe* Camel Detachment preparing to dismount.

## NCO Rank Insignia for the *Feldrock*

The insignia consisted of chevrons of silver braid with straight upward angles. These chevrons were sewn onto a backing of cornflower blue cloth and attached to the left sleeve by means of three hooks. The NCO ranks were indicated by the number of chevrons on the left sleeve of the field tunic:

**Sergeant Major**
(*Feldwebel*) : four chevrons

**First Sergeant**
(*Vize-Feldwebel*) : three chevrons

**Sergeant**
(*Sergeant*) : two chevrons

**Corporal**
(*Unteroffizier*) : one chevron

**Lance Corporal**
(*Gefreiter*) : the insignia of the Lance Corporal was a small decorative button on each side of the tunic collar, made of metal and displaying the heraldic Imperial eagle.

The **Kordfeldrock** (originally called the *Kordlitewka*) was a tunic also made of velvet-like ribbed Manchester cord, but was similar in cut to the Home *Litewka*. It had breast pockets and side pockets with simple flaps without buttons, a concealed front, and no collar patches. Examples can be found without breast pockets. The *Kordfeldrock* was intended to replace the *Kordwaffenrock* and used the same rank system as the *Feldrock*.

## Rank and File (*Mannschaften*)

The hatband and edging were in cornflower blue wool ribbon. The cap was the same as for NCOs. Shoulder straps had black/white/red cords, metal buttons and cornflower blue piping. As on the *Feldrock*, the insignia for lance corporal (*Gefreiter*) was a decorative button on each side of the tunic collar, made of metal and displaying the heraldic Imperial eagle.

*Officers and men of the Schutztruppe at the Neues Palais, Potsdam, 15 June 1894.*

## Musicians (*Spielleute*)

The insignia of musicians consisted of so-called **swallows' nests** (*Schwalbennest*) of cornflower blue cloth, attached to the shoulders by means of four hooks. The trimming was made of white wool (or silver braid for NCOs) and consisted of one horizontal and seven oblique bars running from the back to the front border.

## Band Leader (*Musikleiter*)

The band leader also had a 7 cm silver fringe on the swallows' nests. According to regulations for mounted troops, the striped bars should have been oblique, but after 1914 were usually worn straight. The trumpet cord was in black/white/red.

## Marksmanship Insignia
(*Schützenabzeichen*)

The marksmanship insignia was the same as the Prussian army.

### Gun Layer (*Richtkanonier*)

The insignia for the artillery gun layer consisted of a yellow three-flamed grenade, worn on the left lower sleeve.

### Officer's Shoulder Straps

Epaulettes were not worn by the *Schutztruppe*. Officer's shoulder straps on all uniforms followed the pattern of the Marine Infantry, but without the Imperial crown badge. They were made of silver silk with black and red threads on a cornflower blue background. *Schutztruppe* artillery officers wore the bursting bomb badge of the Prussian field artillery. Officers' buttons on the shoulder straps and tunic were of silvered metal.

### Trousers, Overcoat and Boots

The **Field Trousers** (*Feldhose*) were also made of khaki drill, with cornflower blue piping down the outer seam. No puttees, gaiters, or leggings were worn with the trousers. The **Riding Trousers** (*Kordreithose*) were made of the same corduroy as the *Kordwaffenrock* and had no piping.

The **Greatcoat** (*Mantel*) was a long riding coat with a wide turn-down collar made of light grey cloth. Other ranks wore a blue collar patch with short white braid (*Litzen*) in the style of the Prussian grenadiers. NCO insignia consisted of a narrow white strip above the *Litzen*, with the *Feldwebel* having two stripes. The coat had cornflower blue shoulder straps held with white metal buttons bearing the Imperial crown. The turn-down collar had two hooks with corresponding eyelets to

Khaki *Feldrock* Uniform of a German South-West African *Schutztruppe* Soldier, including Boots, Spurs and *Südwester* Hat.

His Field Equipment consists of brown leather Cartridge Belt, Canteen, *kS98* Bayonet with Troddel, and *Gew M98* Rifle.

fasten it. There was also a hook at the top left front which fastened under the collar with an eyelet if needed. The front of the coat was fastened with six white metal buttons bearing the Imperial crown. In line with the last button were two slightly sloping pockets without flaps. The rear skirt had a 60 cm-long slash, which could be closed by two concealed buttons. At the rear of the waist was a two-part cloth belt, fastened by a white metal button bearing the Imperial crown. Officers wore the double-breasted **Riding Coat** (*Paletot*) of the army with a cornflower blue collar, with shoulder straps added in 1903. Officers often wore the riding coat in the field.

### Boots (*Stiefel*)

Made of natural brown leather, the boots had a straight shaft up to the knee and a 4 to 5 cm-wide spur strap. The spur was horizontal, 3 cm long, with a spur wheel at the end. Some boots had marching studs in the soles, others did not.

### *Südwester* Hat

The signature *Schutztruppe* hat was known as the **Südwester**. Made of grey felt, it had an oval head 15 cm tall, with a 3.5 cm-wide band of cornflower blue ribbon, with a flat bow on the left side. The brim of the hat was 12 cm wide and had a cornflower blue ribbon edging. The ribbon and edging often bleached out over time. Officers' hats were high quality felt with silk ribbon, whereas NCOs' and other ranks' hats were generally of lesser quality. The right side of the brim was turned up and

*Schutztruppe* Soldier in the khaki *Feldrock* with the South-West Africa Campaign Medal with a Battle Clasp.

*Schutztruppe Gefreiter* in Corduroy *Kordrock*.

held vertically by a 5.5 cm-diameter Imperial cockade. The cockades of officers and senior NCOs were more deeply embossed and painted in black/silver/red. Junior NCOs and other ranks wore a flatter cockade painted in black/white/red. The brim was further fixed by a brass push button between the hat and its brim. Inside, the hat was equipped with a silk lining which was often lost very quickly from wear. Some hats had two small ventilation holes on the left side.

Right side of *Schutztruppe Südwester* Hat for Other Ranks in German South-West Africa.

*Schutztruppe* Soldiers before their departure for German South-West Africa.

*Schutztruppe* Soldier wearing the *Feldrock* with Corduroy Riding Trousers.

Left side of *Schutztruppe Südwester* Hat for Other Ranks in German South-West Africa.

**Field Cap**

The field cap (*Mütze*) was of Prussian army style, but was made of sand brown corduroy. The cap's hatband and piping around the brim were of cornflower blue cloth with an Imperial cockade on the front of the hatband. The cap had a black leather peak for all ranks (the usual peakless *Krätzchen* of the army was not worn by the *Schutztruppe*). From 1893 to 1895 a low Kepi was worn with cornflower blue hatband and piping, an Imperial cockade and a square black leather peak.

Two Tropical Helmets of the
South-West African *Schutztruppe*.
These Helmets were authorized for use in
German South-West Africa in 1891, but were deemed
impractical and soon retired. L to R: Officer and NCO models.

*kS98 Schutztruppe* Bayonet.

*German South-West Africa* 31

*

German South-West African *Schutztruppe* Officer's Home Uniform and Greatcoat worn by *Oberleutnant* Daubenkopf.

Trumpet with Swallows' Nest for the shoulder of the Tunic.

Photos [*]: Fotoarchiv Firma Helmut Weitze Militärische Antiquitäten KG

Greatcoat for a *Schutztruppe* NCO with Cartridge Belt (3rd model), long *S98* Bayonet, and Canteen.

*Schutztruppe* Rider, 30 mm flat figure, and dismounted Rider in Greatcoat, 54 mm, both painted by the author.

Two studio portraits of *Schutztruppe* Soldiers.

Rank Insignia of a *Sergeant*.

ID Tag for a *Schutztruppe* Soldier.

*Schutztruppe* Soldier,
54 mm figure painted by the author.

*Schutztruppe* Bugler,
54 mm figure painted by the author.

Uniform plate from the early 1890s, showing the *Schutztruppe* for German East Africa on the left and German South-West Africa on the right.

Buttons of the *Schutztruppe*. The top and bottom rows are *Schutztruppe* Tunic Buttons with the Imperial Crown. In the centre of the middle row is a large Collar Button worn by both *Schutztruppe Sergeante*n and *Feldwebel* in white metal with the Imperial Eagle. Seen on its left and right are smaller yellow metal Collar Buttons of an *Ober-Lazarettgehilfe*.

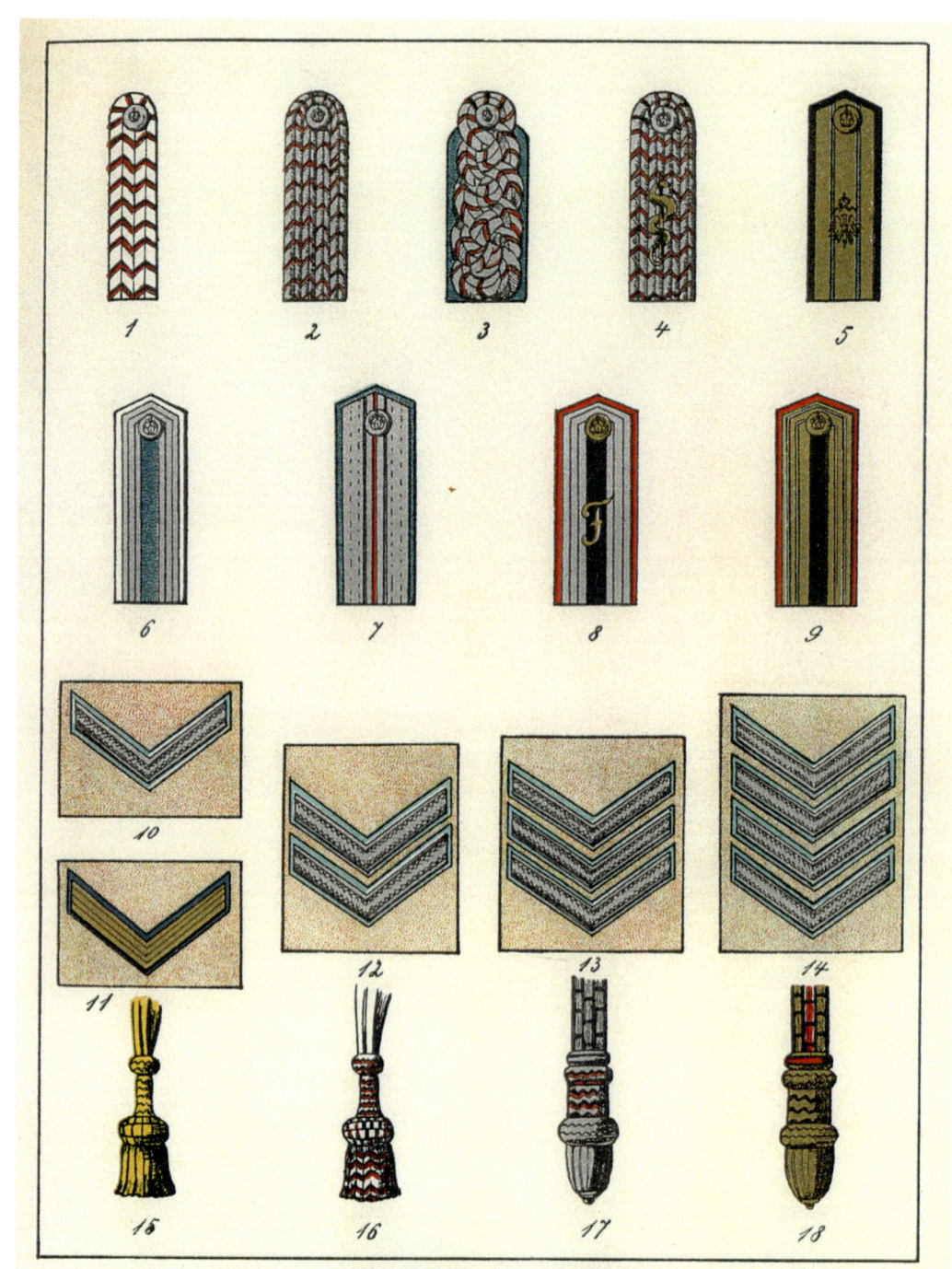

**Deutsche Schutztruppen in Afrika.**

**Achselstücke.** 1 Achselschnur der Mannschaften, 2 Achselstück der Subaltern-Offiziere, 3 Achselstück der Majore, 4 Achselstück der Ärzte, 5 Achselstück der Roßärzte, 6 Achselstück der Zahlmeister-Aspiranten, 7 Achselstück der Zahlmeister, 8 Achselstück der Oberfeuerwerker, 9 Achselstück der Oberbüchsenmacher. **Rangabzeichen am Dreßrock, am Oberarm.** 10 Unteroffizier, 11 Lazarettgehilfe, 12 Sergeant, 13 Vize-Feldwebel, 14 Feldwebel. **Portepees.** 15. Portepee des Unter-Büchsenmachers, 16 Portepee der Unteroffiziere, 17 Portepee der Offiziere, 18 Portepee der Ober-Büchsenmacher.

Shoulder Straps, Rank Insignia and Sword Knots.

*German South-West Africa* 37

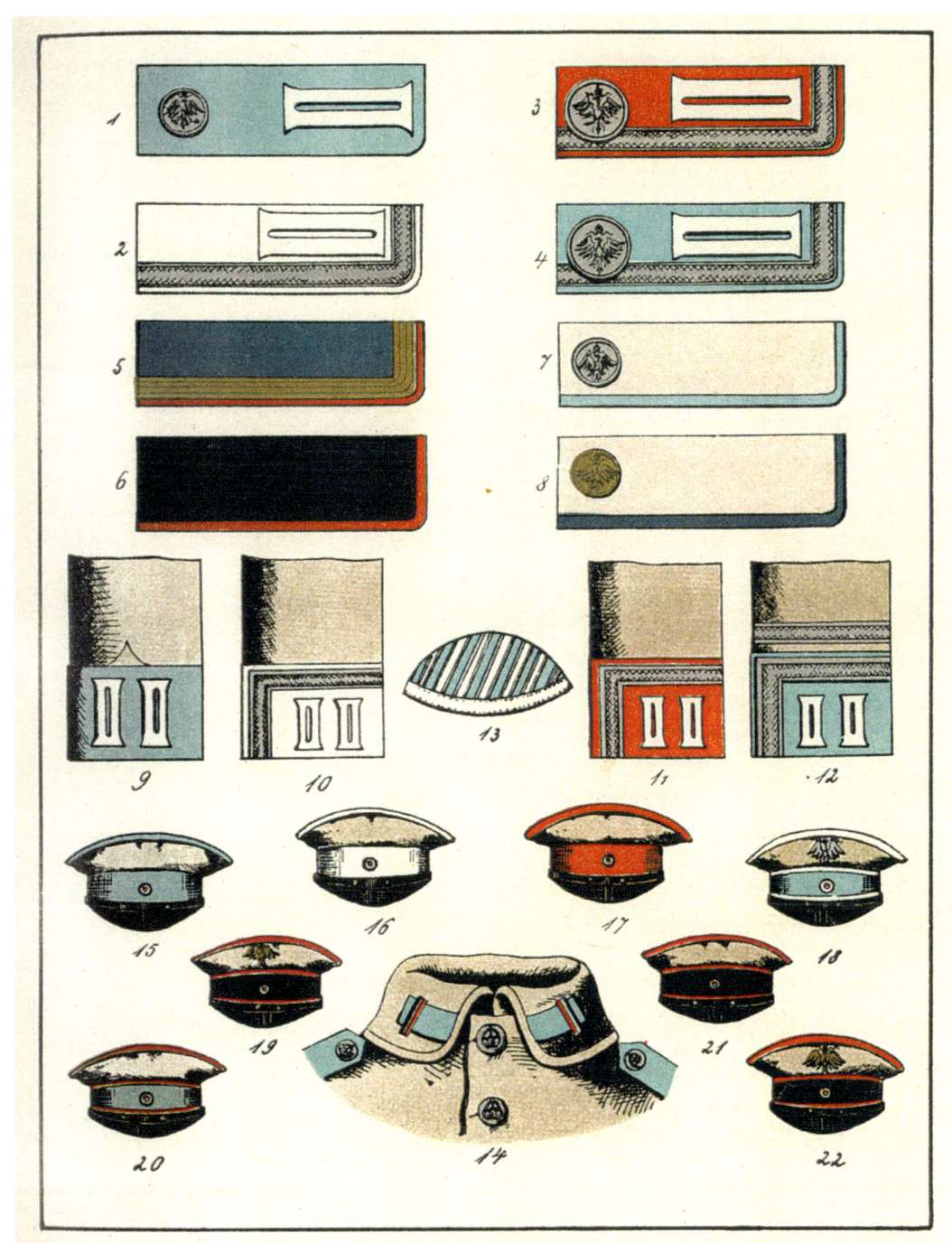

**Abzeichen der Schutztruppe.**

1 Kragen der Gefreiten (Südwestafrika). 2 Kragen der Unteroffiziere (Ostafrika). 3 Kragen der Sergeanten (Kamerun und Togo). 4 Kragen der Feldwebel (Südwestafrika). 5 Kragen der Lazarettgehilfen. 6 Kragen der Unter-Büchsenmacher. 7 Kragen der Gefreiten (Südwestafrika). 8 Kragen der Unter-Lazarettgehilfen. **Ärmelaufschläge.** 9 Gefreiter (Südwest-Afrika). 10 Unteroffizier (Ost-Afrika). 11 Sergeant (Kamerun und Togo). 12 Feldwebel (Südwest-Afrika). 13 Schwalbennest der Hornisten. 14 Mantelkragen der Unteroffiziere. **Mützen.** 15 Südwestafrika. 16 Ostafrika. 17 Kamerun und Togo. 18 Zahlmeister. 19 Roßarzt. 20 Ärzte. 21 Oberfeuerwerker. 22 Büchsenmacher.

Insignia of the *Schutztruppe*.

Peaked Cap for the Corduroy Uniform of a *Veterinäroffizier*.

## *Schutztruppe* Officials

The non-combatant specialist officers of the *Schutztruppe* wore the same uniforms as the *Schutztruppe* but with identifying insignia for their specialist trades. All officials assigned to the *Schutztruppe* were classed as military personnel, even though some of their jobs were considered civilian in the Prussian army. As such all wore uniforms and carried the golden sword knot (*Portepee*), although not all would have done so in the Prussian army.

## Medical Officer (*Sanitätsoffizier*)

The *Südwester* hat band and the cap band were of dark blue velvet piped on the top and bottom with a line of ponceau red cloth. The silver cords of the shoulder straps of the medical officers with staff officer rank were interspersed with black/white/red threads. Assistant, senior, and staff doctors had an additional 1 mm-wide, black silk twisted cord between the silver/black/white/red cords of the shoulder. For medical officers of staff rank this was a 3 mm-wide, black silk-edged cord. The base of the shoulder strap was made of dark blue velvet, and on the straps was a gilt Aesculapius rod: a snake twisted around a rod. The tunic had gilded buttons and was piped in dark blue.

## Veterinary Officer (*Veterinäroffizier*)

The *Südwester* hat band and edging was of black ribbon. The cap band was of black cloth bordered by crimson piping above and below. On the front of the cap was a small gilt heraldic Imperial eagle above the cap band. The shoulder straps were like those of medical officers, only having a snake badge without the staff. The backing was of black cloth. Buttons on the tunic were gilded and the piping was black.

## Ordnance Officer (*Feuerwerksoffizier*)

The *Südwester* hat band was of black velvet. The cap band was also made of black velvet with piping in ponceau red cloth on the top and bottom edges, the same piping also along the cap edge. The shoulder straps were covered with black velvet and bore a gold-plated "F." The tunic had gilded buttons and was piped in black.

## Ordnance NCO (*Ober-Feuerwerker*)

The *Ober-Feuerwerker* held the rank of *Feldwebel*, and had black cloth trimmings on the hat and black wool on the cap until the end of 1913. After that they had black cloth

Junior Paymasters (*Zahlmeister-Aspiranten*) in Corduroy Uniform. In 1906 their rank title changed to *Unter-Zahlmeister*.

with red piping. The shoulder straps were red in the middle, black at the edge with silver braid and a gilded "F" badge. The tunic had brass buttons and was piped in black.

**Paymaster** (*Zahlmeister*)

The hatband was of dark blue silk ribbon. The cap band was in dark blue cloth, piped at the top and bottom in white, with the top edge of the cap also piped in white. At the front of the cap, above the cap band, was a small silver heraldic Imperial eagle. The shoulder strap backing was of dark blue cloth. The tunic had white metal buttons and was piped in dark blue.

**Junior Paymaster** (*Unter-Zahlmeister*)

The hatband was of dark blue silk ribbon. The cap band was also in dark blue with piping in white as for the Paymaster, but without the heraldic eagle. The tunic had white metal buttons and was piped in dark blue. Until the end of 1913 dark-blue shoulder straps with white piping were worn. On the outer edge of the shoulder straps was a silver band with two black longitudinal threads. From the end of 1913 onwards, white shoulder straps trimmed with silver and blue braid were worn.

**Senior Gunsmith** (*Ober-Büchsenmacher*)

The hatband was of black silk ribbon.

Military Pass for *Schutztruppe* Personnel.

The cap band was also in black cloth, and both were piped at the top and bottom in ponceau red. The shoulder strap was of black cloth with ponceau red piping. A gold lace braid was sewn around the outer edge of the shoulder strap, with blue vertical stripes. The tunic had gilded buttons and was piped in black.

### Junior Gunsmith (*Unter-Büchsenmacher*)

The hatband was of black wool ribbon. The cap was as for the Senior Armourer, but a small bronze Imperial eagle was worn above the cap band. NCOs' aiguillettes were not worn on the field tunic. The tunic had yellow metal buttons and black piping.

### Hospital Assistant (*Lazarettgehilfen*)

The hatband was in dark blue wool ribbon. The cap band was also in dark blue cloth, piped at the top and bottom in ponceau red as well as along the rim of the cap. NCOs wore the aguillette cords on their tunics with buttons in yellow metal and dark blue piping. Rank insignia was shown as gold braid chevrons.

### Military Construction Official (*Militärbausekretär*)

Construction officials had dark blue insignia with crimson cloth piping on the cap and tunic, silvered Imperial crown buttons, and a silvered Imperial eagle shield badge on the hat and cap. Shoulder straps were of gold and dark blue interlaced flat cords, atop a dark blue cloth backing. After six years' service a silvered rosette was added to the shoulder straps, below the silvered Imperial eagle shield badge.

*Schutztruppe* cleaning their Rifles. Note *MG01* Machine Gun.

### Weapons Inspector (*Waffenrevisor*)

Weapons Inspectors had black insignia with ponceau red cloth piping. Officials had golden shoulder straps and two gilded crossed rifles above a gilded official's shield badge.

### Armourer (*Waffenmeister*)

The armourer had black insignia with red piping and red shoulder straps with golden

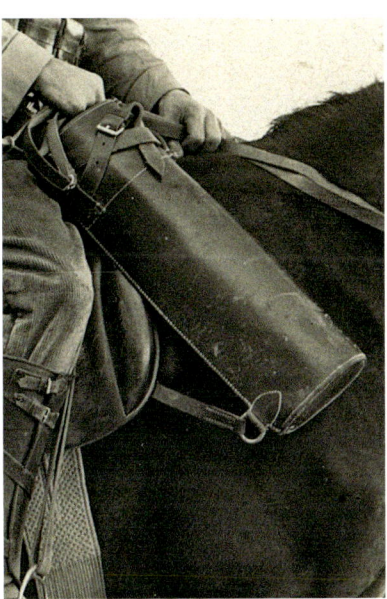

*Schutztruppe* Rider on horseback (with Rifle Bucket attached to right front of Saddle).

braid. Atop that was a gilded official's shield with crossed rifles, or, in the case of the artillery, crossed cannon barrels. The greatcoat collar was also black with red piping.

### Dentist (*Zahnarzt*)

The dentist of the South-West African *Schutztruppe* had ponceau red insignia on his hat, cap and tunic. The shoulder straps were silver on blue backing cloth, with an official's shield badge. Buttons were silvered and bore the Imperial crown.

### Staff Pharmacist (*Stabsapotheker*)

The staff pharmacist had crimson insignia on his hat, cap, and tunic, with silver *Litzen* on the collar and on the Swedish cuffs. The shoulder straps were of silver, blue, and crimson, with a gold official's shield badge and gilded rosettes.

### Munitions Officer (*Magazinmeister*)

The munitions officer had dark blue insignia with yellow piping, gilded Imperial crown buttons, and yellow shoulder straps with gold braid, edged in dark blue.

### Saddler (*Sattler*)

The saddler had black insignia with white piping, the same shoulder straps and braid as the munitions officer, and gilded buttons. They wore the other ranks' greatcoat with dark blue or black collar patches with the shoulder straps of the tunic.

### Chaplain (*Militärgeistlicher*)

The military chaplain wore the field uniform without shoulder straps, and with violet insignia on the hat and cap. A white cross was worn on the front of the hat and on the cap above the Imperial cockade.

### Medical Other Ranks (*Sanitätsmannschaften*)

The medical rank and file wore a yellow Aesculapius badge on the upper right sleeve.

Studio photo of *Schutztruppe* Rider and Mount.

*Oberstleutnant* Heydebreck (Commander of the *Schutztruppe* for German South-West Africa at the outbreak of the First World War).

*Offiziersarzt* on horseback.

Rifle Bucket.

# German South-West Africa

## Equipment of the *Schutztruppe*

### Cartridge Belt (*Patronengürtel*)

To equip the *Schutztruppe* a special cartridge belt was introduced, one not commonly used by the army. To carry the weight of 120 cartridges, a special harness made of strong brown cowhide was designed, which held twelve leather pouches, each containing ten cartridges. On the left side of the belt was a frog for the **Bayonet** (*Seitengewehr*). On the right side was a steel D-ring for the **Canteen** (*Feldflasche*), which was made of aluminum and covered with felt. The shoulder straps crossed in the back and ensured that the considerable weight of the equipment was not borne on the belt alone. The belt buckle was of open design, located at the rear.

Over time there were three models of the cartridge belt. The 1st model had ten cartridge pouches, the flaps of which were fastened with a leather loop. The 2nd model added an additional cartridge pouch to the front of each shoulder strap, with flaps still fastened with leather loops, bringing the total number of cartridge pouches to twelve. The 3rd model retained the twelve cartridge pockets with leather loops but added press studs to each flap, to prevent cartridge strips from falling out.

### Riding Equipment

Each soldier had the grey greatcoat strapped behind the **Cavalry Saddle** (*Kavalleriesattel*) under a waterproof tarpaulin of brown canvas. Below the Saddle on either side were spacious **Saddle Bags** (*Packtaschen*), each 25 cm tall and 20 cm wide. One bag had a buckle on top, the other a leather loop. A steel ring attached to the rear wall of each bag

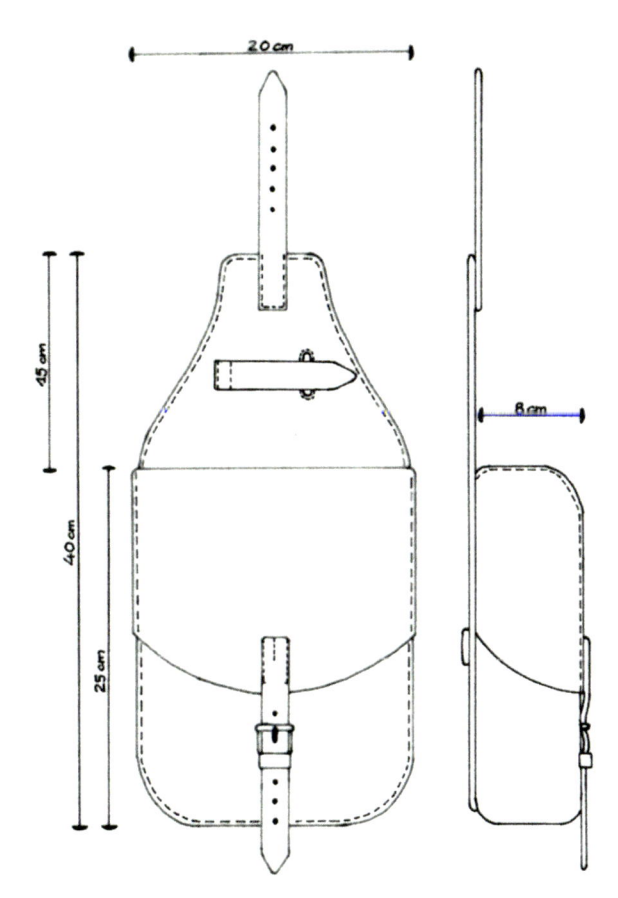

*Schutztruppe* Saddle Bag, drawing by the author.

Cavalry Saddle as used in German South-West Africa.

Saddle Bags of a *Schutztruppe* Rider.

was for hanging a **Water Bag** (*Wassersack*). Saddle bags without the steel ring can also be found. The water bags were square (30 x 30 cm), made of waterproof hemp fabric, and each held three liters. They had short wooden syringe-like tubes in each upper corner which were closed by a cork attached to a leather cord. A folding water bucket also made of hemp was used in conjunction with a Verkefeld filter, and carried in one of the two saddle bags. The **Rifle Bucket** (*Gewehrschuh*) was about a metre long and was attached at the front right of the saddle. The loaded rifle with safety catch on was carried in the bucket muzzle upwards with a loose carrying strap, but was itself unstrapped and free for immediate use. The rifle used was the infantry *Gewehr M98* rifle with convertible sight.

At the left of the saddle was the **Cookware** (*Kochgeschirr*), held in place inside a series of leather straps with a buckle on the bottom. The **Bridle** (*Zaumzeug*) consisted of an S-curb bit and halter. The saddle was similar to that used by the cavalry in the German army with an additional felt saddle pad. The 7th Camel Mounted Company carried the rifle bucket on the rear right side of the saddle, and their saddle bags were of a taller, narrower shape.

### Bayonet Knot (*Troddel*)

The bayonet knot for sergeants and junior NCOs was black/white/red. That of other ranks followed the usual army pattern.

### Identification Tag (*Erkennungsmarke*)

The identification tag was made of aluminium, 5 cm x 3.5 cm, with two small holes at the top to attach a lanyard. Stamped onto each tag was: "*Schutztr SWA*" and the soldier's number. The tag also came in other shapes and sizes.

### Other Ranks' Chest (*Mannschaftskoffer*)

Every man had his own tropical chest, made of highly galvanized sheet iron with double brass locks and a brass carrying handle. It was 45 cm long, 29 cm wide, and 10 cm tall.

Water Bag.

German South-West Africa 45

Horse Bucket made of hemp.

Rifle Bucket.

Schutztruppe Rider Mess Kit.

Horse Bridle.

Camel Rider of the 7th Company (Gochas),
54 mm figure painted by the author.

Patrol mounted on Camels.

Saddle bags for Camel Riders were narrow and about 45 cm tall.

Service Log Book for a Civilian Official serving with the *Schutztruppe*.

*Schutztruppe* Other Ranks' Chest.

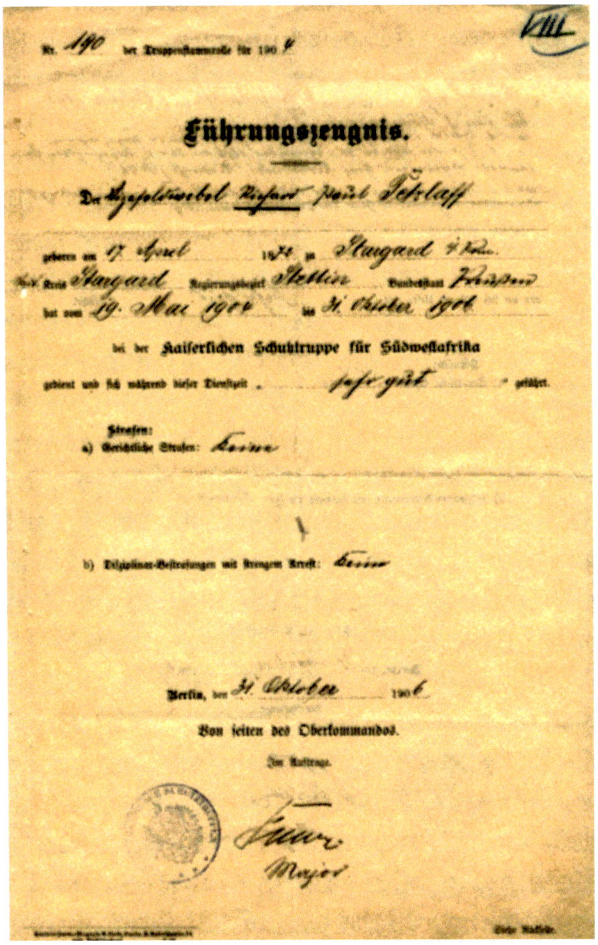

Service Record of a Vice Sergeant Major in the *Schutztruppe*.

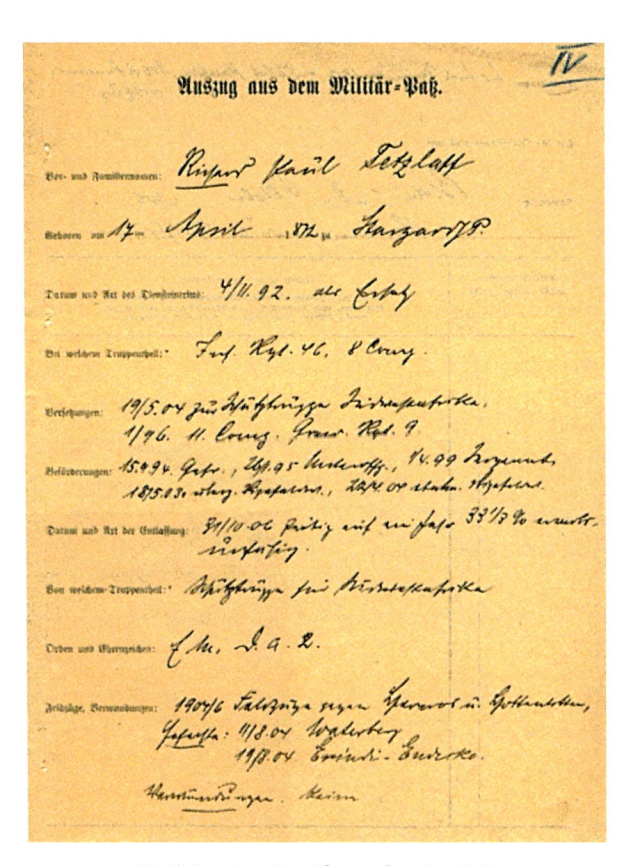

Civil Service Certificate for the *Schutztruppe*.

Field Postcard from German South-West Africa.

Postcards from German South-West Africa.

## Weapons of the *Schutztruppe*

### Rifles and Carbines

From 1884 to 1889, the **Gewehr M71** with a calibre of 11.2 mm was commonly carried. In many cases, however, the light infantry **Jägerbüchse M71** was used. Some of the *Jägerbüchse* rifles were equipped with a special multi-loading magazine. It was inserted into the breech sleeve from below and extended around the shaft of the rifle to the right. Starting in 1889, the **Gewehr M88** and **Karabiner M88** carbine with a calibre of 8 mm were introduced to the *Schutztruppe*. It was with these weapons that the *Schutztruppe* were armed during the uprisings of the 1890s – weapons which were still in use during the great rebellions in the years 1903 – 08. In 1904, the first **Gewehr M98** rifles arrived for the *Schutztruppe*. These were the normal versions of the rifle with a straight bolt, still set up for the *Gewehr M88* 8 mm cartridge. In the autumn of 1908, existing rifles began to be modified to take the new S-cartridge. This was completed by 1 April 1909.

The first artillery battery was established in 1894 and was armed with the *Karabiner M88* carbine. The machine gun unit and the artillery reinforcements in the years 1904 – 05 brought with them various versions of the **Karabiner M98** from the homeland. The first version of the *Karabiner M98* had a barrel length of 43.5 cm and was set up for the *M88* cartridge. At this time, it was used to arm the artillery and machine gun units, until the introduction of the S-cartridge for the short *Karabiner M98*.

*Schutztruppe* Machine Gun Crew in firing position.

Towards the end of 1908, the *Schutztruppe* carried out tests on the long version of the *Karabiner M98* with the S-cartridge. This extended carbine with a 74 cm-long barrel came to be called the **Karabiner 98a**, and would eventually be introduced to the entire force. It was from this weapon that the **Schutztruppen-Gewehr M1898** was developed, after armed trials.

### Schutztruppen-Gewehr M98

This rifle was developed especially for the needs of the *Schutztruppe* in German South-West Africa. It had a barrel length of 74 cm and was set up for the S-cartridge. The muzzle sight was the same as the normal type of *GewM98*. The sights were the hitherto long sights but with a 200-metre position as the nearest sighting point rather than 400-metre, as was standard for the S-cartridge *GewM98*. With this setting, close opponents in the thick bush could be easily targeted. The maximum sight setting was as usual 2,000 metres. Like all long-sights, it was possible to change the sighting position at points up or down in 100-metre increments by pressing one of the two side lugs. By pushing the lugs, the sight slider could be pushed forward or backward for rough adjustments. This special type of 200-metre sight on the *Schutztruppen-Gewehr M98* was not used in any other weapon using the S-cartridge.

The bolt lever was bent in the semi-circular manner of the carbine, the ball head flattened on one side, with a fish-skin patterned imprint. The stock corresponded to that of the standard *Gewehr M98* rifle, with a lengthwise recess in the shaft to accommodate the curved bolt lever. The rifle strap was also like that of the *Gewehr M98*, in that it was not attached at the side (as in the case of the carbine), but at the front, because the *Schutztruppe* were essentially mounted infantry rather than cavalry. The stamp plate in the butt bore the imprint "KS" (*Kaiserliche Schutztruppe*) with the troop number. Below that was an "S", denoting the weapon's conversion to S-calibre.

### Bayonets

From 1884 to 1904 the **Seitengewehr M71/84** bayonet was part of the basic equipment. In 1904, reinforcements brought with them the long **Seitengewehr M98** bayonet. Artillery and machine gun units had the short **kS98** bayonet, which was introduced for the entire force starting in 1909. This bayonet had a blade 25 cm long and was provided with a saw edge on its spine. The handles were made of riveted leather with embossed fish-skin pattern grips. The head of the handle had the so-called bird head design. The fitting of this bayonet was in the usual manner for the *Gewehr M98* rifle.

### Pistols

The equipment of the *Schutztruppe* originally included the **Reichsrevolver M83**. Of these revolvers, around 500 were still available in 1914. Starting in 1904 during the Rebellions, Parabellum pistols arrived with a calibre of 9 mm and a barrel length of 15 cm. Around 1910 the normal **Parabellum 9 mm Pistole M08** arrived in South-West Africa. Before that, many

Above and opposing page: the Erhardt 7.5 cm Quick-Firing Recoilless Mountain Gun, dismantled and carried by Mules.

Mauser 7.63 mm pistols with shoulder stock were carried. The Browning 7.65 cm *M1900* pistol was also very popular among the officer corps. Both the Mauser and the Browning guns were usually privately purchased weapons.

## Machine Guns (*Maschinengewehre*)

The first machine gun which arrived for the *Schutztruppe* in 1894 had a calibre of 11.2 mm. It was likely issued on an experimental basis as part of the *Schutztruppe* reinforcements necessitated by the First Nama Rebellion. No documentation about its origin, type designation or duration of service exists. The only proof is a photograph belonging to a member of the troop, who credibly stated that this particular machine gun was in use by the *Schutztruppe*. Based on several other statements, the calibre specification of 11.2 mm can be assumed with certainty. Whether this round was same as the Mauser cartridge *M71* or *71/84* cannot be determined. According to authentic reports, the SMS *Habicht* brought one machine gun for the *Schutztruppe* at the beginning of the Herero Uprising. Further reinforcements brought seventeen more machine guns. Starting in 1909 all machine guns were uniformly converted to take the S-cartridge.

## Artillery (*Geschütze*)

Before the Herero Uprising, the *Schutztruppe* had five **Light Field Guns** (*leichte Feldkanone 73*) transferred from the homeland between 1894 – 96, four **Quick-Firing Guns** (*Schnellfeuergeschützen* 5.7 cm), transferred from home in 1894, and five **Mountain Guns** (*Gebirgsgeschützen* 6 cm), transferred between 1897 – 1903. During the uprising more guns were added; three **Revolver Cannons** (*Revolverkanonen* 3.7 cm) from SMS *Habicht*, eight **Machine Cannons** (*Maschinenkanonen* 3.7 cm) from the Navy Expeditionary Corps, two **Field Guns** (*Feldkanonen 91/93*) transferred from Cameroon, 30 **Field Guns** (*Feldkanonen 96*, field artillery unit and depot), four **Light Field Howitzers** (*leichte Feldhaubitzen 98*, no longer in use after the rebellion), and six **Mountain Guns** (*Gebirgsgeschützen* 7 cm).

In 1908 Erhardt introduced its 7.5 cm **Mountain Gun**, with a recoilless barrel system and protective shield. That same year four *FK96* field guns were returned to Germany and retrofitted with both these improvements. Of the three batteries in existence on 1 October 1908, two were equipped with mountain guns, and one with field cannons.

The 6th Battery with the new Erhardt 7.5 cm Quick-Firing Recoilless Mountain Gun.

Above: 2nd Battery with Erhardt 7.5 cm Recoilless Mountain Gun at Sandfontein, 1914.  Below: Artillery in Windhoek.

## Organisation of the *Schutztruppe* in 1914

In 1914 the total strength of the *Schutztruppe* in German South-West Africa was between 1,600 and 1,800 men. The *Schutztruppe* headquarters was in the capital of Windhoek and from there commanded two districts, the Northern based at Windhoek and the Southern based at Keetmanshoop.

### Northern District (*Nordbezirks*) at Windhoek

- 1st Company at Regenstein and Seeis
- 4th (Machine Gun) Company at Okanjande
- 6th Company at Outjo and Ovate
- 2nd (Mountain) Artillery Battery at Johann-Albrechtshöhe
- Transport train and provisions office at Karibib
- Horse depot at Okawayo
- Artillery and train depot, military hospital, main medical depot, clothing depot, garrison, construction administration and local command at Windhoek,
- Local command and provisions office at Swakopmund.

### Southern District (*Südbezirks*) at Keetmannshoop

- 2nd Company at Ukamas
- 3rd Company at Kanus
- 5th (machine Gun) Company at Chamis and Churutabis
- 7th Camel Rider Company at Gochas and Arahoab
- 8th Company at Warmbad
- 9th Company at Kabus
- 1st Artillery Battery in Narubis
- 3rd. Artillery Battery at Kranzplatz near Gibeon
- Transport Train at Keetmanshoop
- Artillery and train depot, military hospital, medical depot, clothing depot, provisions office, garrison and administration at Keetmanshoop
- Horse depot at Aus
- Camel stud farm at Kalkfontein
- Local command and provisions office at Lüderitzbucht.

Each of these companies had two machine guns. Three companies used pack animals to carry them, while the others used two-wheeled carriages.

In addition to the twelve modern mountain guns of the three batteries, there was an arsenal of 50 old guns of ten different types, dating from the time of the Herero Uprising in 1904. From these *C73* field guns, field howitzers, quick-firing guns and machine and revolver cannons, seven batteries and four half-batteries were newly formed, but they were at a disadvantage against a motorized opponent. Signals, radio and telephone equipment was relatively abundant, and the colony had two aircraft of old design.

From this point of view, the *Schutztruppe* entered the First World War unprepared to face a modern opponent. Reservists were called up, doubling the size of the *Schutztruppe* with three reserve companies and two field batteries equipped with *FK96aA* field guns.

As further reinforcements, on 8 August 1914 a company of Rehoboth-Basters (150 men) and a company of Berseba-Nama (70 men) were formed. Both companies were mounted but existed only until 18 April 1915. For the same purpose, in October 1914 a Cameroonian company was formed for the *Schutztruppe*. These were African soldiers of the Cameroon *Schutztruppe* who had mutinied, and whose punishment was deportation to South-West Africa. When the war broke out, they were formed into a small unit, a portion of which was mounted on oxen. They were disbanded on 24 March 1915, with their soldiers then being employed in police service or on manual working tasks. They eventually returned home after the war.

## Formation of the *Landespolizei*

From 1889 to early 1905 police duties were performed by the *Schutztruppe* and local police authorities. The **Imperial State Police** (*Kaiserlichen Landespolizei*) was formed on 1 March 1905. The *Landespolizei* consisted of civil servants who reported to the colonial governor and not to the *Schutztruppe*. Uniforms were privately purchased, but weapons and equipment were delivered from official *Schutztruppe* stocks.

So long as the colony did not need a strong military force, the modest forces of the *Landespolizei* were deemed sufficient. It was only with the suppression of the Herero and Nama

Postcard from German South-West Africa.

*Schutztruppe* Soldiers mounted on Oxen. This was not common practice.

Rebellions and the subsequent departure of most of the *Schutztruppe* that the civil administration once again exercised its rights and proceeded with a substantial increase in the *Landespolizei*. The personnel were now no longer drawn mostly from the *Schutztruppe*. The colony's budget for 1907 provided funds for 648 **Police Sergeants** (*Polizeisergeant*) and 72 **Senior Sergeants** (*Wachtmeister*), to serve under a staff officer and five **First Lieutenants** (*Oberleutnant*) as police inspectors. On 1 August 1907, *Major* von Heydebreck of the *Schutztruppe* was appointed **Chief Inspector** (*Polizeiinspekteur*) of the *Landespolizei*.

**The Organisation of the Landespolizei was now based on:**

1. The Organisational Rules of the Governor of 1 March 1905.

2. The Memorandum to the Second Supplementary Budget of 1907.

3. The Imperial Order of 4 October 1907.

4. The colonial Governor's Terms of Acceptance for the hiring and distribution of clothing and equipment.

## Uniforms of the *Landespolizei*

The **Hat** was in the style of the *Schutztruppe Südwester* but was made of brown soft-haired felt. It was oval, 15 cm tall with a 10 cm-wide brim. There was a 6 cm-wide hatband of matching brown cotton fabric. On both sides of the hat, buttons were mounted for turning up the brim, which itself had matching eyelets on both sides. On the turned-up right side there was a gilded Imperial crown badge, 5 cm high and 5 cm wide. In the centre of the hatband at the front was the German cockade, as worn on the officer's cap. The hat had an adjustable, brown leather chin strap.

*Landespolizei Sergeant* Ernst Grah.

Corduroy Cap for a *Landespolizei* Officer or Senior NCO.

The **Cap** was the same shape as that worn by the East Asian occupation forces, but made of brown serge fabric. It had a cap band of green cloth and piping of the same color around the upper edge. On the front of the cap was the German cockade. The peak extended 5 cm at the front and was of black patent leather. The adjustable chin strap was also in black patent leather and was attached by two gilded buttons on both sides of the hat.

The **Service Tunic** (*Dienstrock*) was made of brown jersey in the style of the field tunic of the *Schutztruppe*. The gilded buttons had an Imperial crown, and the front of the tunic was fastened by six such buttons. The collar was of green cloth in the color used for piping by the Prussian army's 8th Cuirassier Regiment (*Kürassierregiment Graf Goßler (Rheinisches) Nr. 8*). It was slightly striped and was held closed by two pairs of hooks and eyes.

The **Sergeant** wore a gold star on either side of the collar. The **Wachtmeister** had two stars diagonally above each other. The **diensttuende Wachtmeister** had three stars, two obliquely on top of each other, the third

*Landespolizei Sergeant* wearing the *Südwester* Hat.

*Landespolizei Sergeant* firing from behind his Horse, which was trained to lie down and provide cover.

positioned in the gap. The shoulder straps of the *Sergeant* were 4 cm wide and of the same green cloth as the collar. For the *Wachtmeister* the shoulder strap had a narrow green-gold braid border. The *diensttuende Wachtmeister* had a gold braid border.

The **Tunic** had a straight cut with two 12 cm slits at the bottom of the side seams. There were sewn-on pleated pockets on either side of the breast and front skirts. All pockets had a pointed flap and a small gilded Imperial crown button. A white **Navy Shirt Collar** was worn under the tunic.

**Trousers** were in the cut of Prussian infantry other ranks, but made of brown jersey with green cloth piping. The **Riding Trousers** were as prescribed for Prussian cavalry other ranks, but made of brown cotton cord with leather lining on the knee.

The **Sabre Belt** (*Säbelkoppel*) was of padded brown natural leather, with a brass buckle in front and a shoulder strap worn across the right shoulder. For undress uniform occasions, the belt could be worn without the shoulder strap.

The **Sword Knot** (*Portepee*) was authorised for all police *Wachtmeister*, Sergeants and former warrant officers as well as for civil servants, but with a gold-embroidered leather strap. The **Saber Tassel** (*Säbeltroddel*) was dark green with silver threads as worn by light infantry NCOs (*Oberjäger*) in the Prussian army but with a leather belt.

**Gloves** were of brown leather as introduced for officers of the army.

Short brown lace-up **Boots** of brown natural leather were worn. Strap-on **Gaiters** were also made of brown natural leather and tightly surrounded the lower leg, reaching from above the ankle to below the knee. **Spurs** were of standard pattern with a narrow natural-coloured leather strap worn over the instep of the boot.

The **Greatcoat** was as authorised for other ranks of the *Schutztruppe*, but had six gilded Imperial crown buttons and green cloth shoulder straps as worn on the tunic. The shoulder straps of the *Oberwachtmeister* had gold lace edging, and those of the *Wachtmeister* had green and gold braid.

## Orders of Dress

**Service Dress** (*Dienstanzug*) was worn for security duties on-site, official notifications and appeals, and in court. It consisted of hat, service tunic, long trousers or breeches with gaiters, belt and buckle, and pistol. For *Oberwachtmeister* and *Wachtmeister* the Saber (*Säbel*) was carried in place of the pistol. For *Polizeimeister* on duty off-site, however, the pistol was carried in addition to the saber. For mounted service dress the hat's chinstrap was always worn down.

**Undress Uniform** (*Kleiner Dienstanzug*) was worn for errands on-site and as a suit with the hat at church. It consisted of hat or cap, service tunic, long trousers or breeches with gaiters, belt and buckle, and sword.

**Patrol Dress** (*Patrouillenanzug*) was worn for off-site duty, military exercises and inspections. It consisted of hat, service tunic, breeches, belt and buckle with bandolier, and carbine.

**Appendix to Uniform Regulations**

1. All police ranks (*Polizeioberwachtmeister, Polizeiwachtmeister, Polizeisergeanten*) and police candidates must wear uniform on and off duty.

2. The wearing of civilian clothes is permitted only for:

a) The officer permanently assigned to the secret police.

b) The other police if they are specifically ordered to carry out an official assignment requiring it, or if they are on leave or sick leave.

3. Civilian clothes must be worn when travelling on non-German ships, on home leave and when abroad. The Colonial Office (*Reichskolonialamt*) and Governor reserve the right to allow the wearing of the uniform in exceptional cases.

4. It does not appear appropriate for subordinate officials who are not members of the secret police to wear civilian clothes. Therefore, in the interests of the service, the approval of the governor must be obtained to do so.

**Final Provisions:**

Until further notice, the following can be worn:

For patrol dress: the corduroy *Litewka* of the *Schutztruppe* with collar stars and shoulder straps of the *Landespolizei*, corduroy trousers without piping and high boots.

For dress and service uniforms: the khaki and white uniforms of the *Schutztruppe* without piping and with collar stars and shoulder straps of the *Landespolizei*.

*Landespolizei Sergeant* Wilhelm Wilhelmi on camelback (Hasuur/Aroab).

# Equipment of the *Landespolizei*

The **Rifle Bucket** was the same in form and method of carrying as that of the *Schutztruppe*. However, in 1909 a new method of carrying the carbine on the saddle was introduced. This was to enable the rider to stay in contact with his weapon, even if he was willingly (or unwillingly) separated from the horse. For this purpose, the carbine was connected by a ring on the rider's belt, just to the rear of the bayonet. A contraption, designed to hold the carbine's barrel in place with a spring bracket whilst riding, was attached to the saddle's girth on the left side at the height of the stirrup. This arrangement undoubtedly offered numerous advantages compared to older designs. If the rider dismounted, he had immediate access to the carbine via his belt. However, through the continuous duress inflicted by the constant use of stirrups and spurs while

Actual size of the Imperial Crown worn on the *Landespolizei Südwester* Hat, drawn by the author.

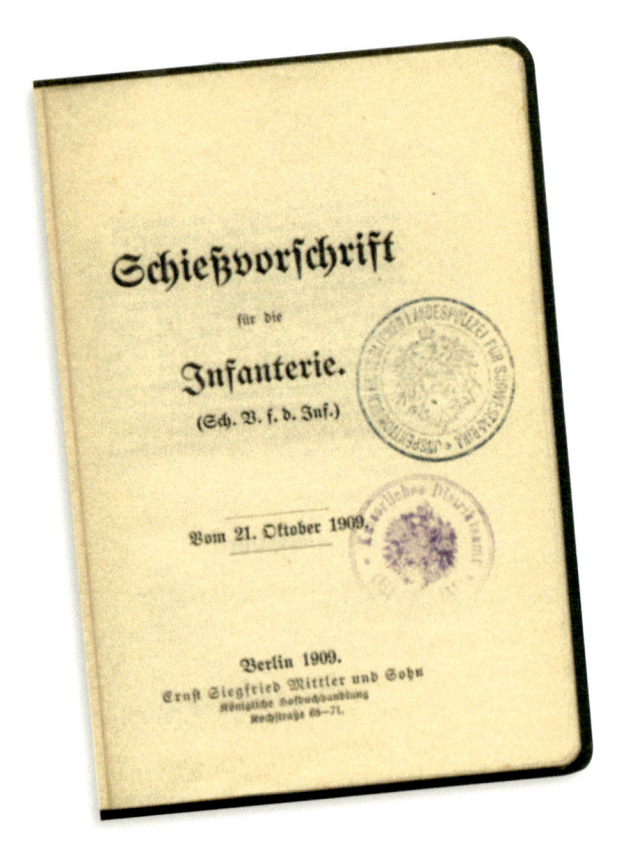

Infantry Shooting Manual with a stamp from the *Landespolizei* Inspector at Omaruru District Office.

trotting or galloping, the wear and tear on the gun's stock and barrel was quite substantial, to the point where the workability of the weapon became questionable. Furthermore, when a rider of shorter stature dismounted he could inadvertently bash the rifle's butt into the ground, with a likely chance of breaking it. Therefore, the *Landespolizei* returned to the old method of carrying the rifle bucket on the right side of the saddle.

The *Landespolizei* **Cartridge Belt** was made of brown cowhide and was worn diagonally over the left or right shoulder. The belt had eight pouches, each holding five rounds. These pouches were closed with brass buttons and when strung together extended to around 50 cm. The pouches had no bottom. The upper part of the pouch was 7 cm tall, and when cartridges were inserted, their tips extended through the bottom by 2.5 cm. The buckle to fasten the strap was located at the end of the string of cartridge pouches.

The **Bridle** and **Halter** were of natural leather with a buckle on the side for attaching the bit. This was a Pelham bit with a hinge curb bit, which was provided with a ring on one side and a snap hook on the other. The **Tether**, like that of the *Schutztruppe*, was made of braided leather.

The **Saddle** was of the latest army pattern, with the frame and seating belt made from one piece and a strap made from hemp.

The **Saddle Bags** were arranged two at the front and one at the rear. Upon introduction of the new rifle carrier, one saddle bag was worn at the front and two at the rear, as done by the *Schutztruppe*.

The **Horse Blanket** was, like that of the *Schutztruppe*, an underlay of brown-green cloth, with a map pocket of natural leather.

## Weapons of the *Landespolizei*

The *Landespolizei* did not carry the straight sword of the *Schutztruppe*. After a trial period, they adopted a hussar-style **Sabre** (*Säbel*) with a curved blade.

Various **Rifles** were used by the *Landespolizei*: *Gewehr M71, Gewehr M88, Gewehr M98, Gewehr M98S, Karabiner M71, Karabiner M88, Karabiner M98* and *Karabiner M98S*.

Several types of **Pistol** were also used: *Revolver 83,* Roth Sauer 7.65 mm automatic, Dreyse pistol and the *Armee-Pistole M08* 9 mm Parabellum.

## African Police Auxiliaries (*Polizeidiener*)

The most important reason for using African policemen was their role in the supervision of the indigenous population. During investigations of offences among the natives they were indispensable. They were also responsible for maintaining order in the Herero villages.

**African Police Auxiliaries** (*Polizeidiener*) wore uniform clothing without any piping or insignia. Their hats were the same colour as those of the German police, with the gilded letters '*LP*' 5 cm high on the right side turned-up brim. The mounted equipment was the same as that of the German *Landespolizei*. As far as their service required, each *Polizeidiener* was also assigned a particular mount (horse or mule) for which he was held responsible.

According to regulations the *Polizeidiener* could not be issued with firearms. Exceptions were made for special purposes – for example, being armed with army revolvers while transporting prisoners. For normal duties the *Polizeidiener* was armed with only the infantry bayonet, carried on a belt with no shoulder belt.

By a decree of 28 February 1912, the rank of **Feldkornett** was tentatively introduced, exclusively for the *Polizeidiener* of the Rehoboth district. These were primarily under

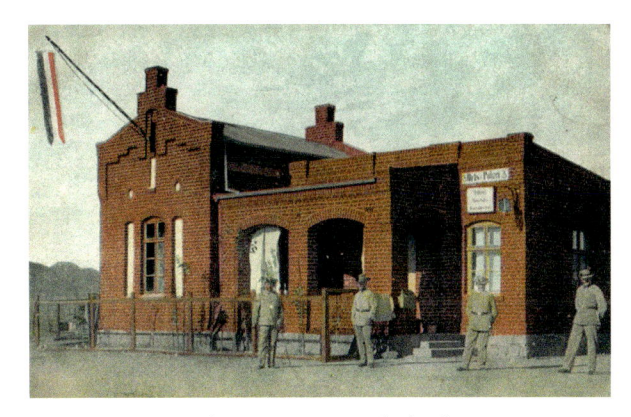

*Landespolizei* Police Station in Rehoboth.

Carbine attached to Ring on Rider's Belt.

Carrying the Rifle in Rifle Bucket.

Corduroy Cap for a *Schutztruppe* NCO.

Postcard touting one brand of *Südwester* Hat.

Other Ranks' Corduroy *Litewka (Kordfeldrock)* for the South-West African *Schutztruppe*.

Corduroy Tunic for a *Schutztruppe* Paymaster.

German South-West Africa

*Landespolizei Südwester* Hat.

Hat Photos: Siegenot Haluschka.

*Landespolizei-Wachmeister* Shoulder Strap

*Dienstrock* for a *Landespolizei-Wachmeister* in German South-West Africa.

Photo courtesy Jan K. Kube Auctions, Sugenheim/Germany.

*Landespolizei* Sword with *Portopee*.

Photo: Siegenot Haluschka.

the command of the district official but were also subordinate to all German police. The *Feldkornetts*' uniform was the same as the *Polizeidiener*, but with shoulder straps and rank insignia on the sleeve. The *Feldkornetts* did not prove useful, however, and they were dismissed in October 1913.

### Organisation of the *Landespolizei* in 1914
- 7 **Officers**
- 9 **Administrative Staff**
- 68 **Senior Police Sergeants** (*Polizeiwachtmeister*)
- 432 **Police Sergeants** (*Polizeisergeanten*)
- 50 **Contracted Policemen** (*Vertragspolizisten*) also **African Auxiliaries** (*Polizeidiener*)

In August 1914, when war broke out in South-West Africa, the tasks of the small police stations grew immeasurably. For weeks, stations on the border were on the front line, facing the enemy. Their tasks were military, and they were incorporated into the *Schutztruppe* with their mounts, equipment and weapons. In their previous role as civil servants, they were not allowed to fight. Now they continued their former duties, but did so as soldiers. The enemy was not familiar with the *Landespolizei's* Organisation – by contrast, their police forces were considered soldiers even in peacetime. Therefore, they counted both *Schutztruppe* and *Landespolizei* as fighting troops, and both were interned after occupation. Captured reservists and *Landwehr* men, by contrast, were released.

African *Polizeidiener*.

## The First World War in German South-West Africa

On 4/5 August 1914, Great Britain's declaration of war on Germany was transmitted via Lome in Togo to Lüderitzbucht in South-West Africa. The following day it was confirmed directly from Nauen via radio telegraph, and at the same time the state of war was announced by Governor Theodor Seitz. On 7 August, general mobilization was declared in the colony.

South African and British forces invaded German territory from the south and also landed at Walvis Bay. On 9 July 1915, at kilometre 500 of the Otavi Railway near Khorab, a cease-fire agreement for the honourable handover for the *Schutztruppe* was signed by Governor Seitz, *Schutztruppe* commander *Oberstleutnant* Franke, and the South African General Louis Botha. Under the terms of the Treaty of Versailles in January 1919, the mandate for German South-West Africa was awarded to the Union of South Africa.

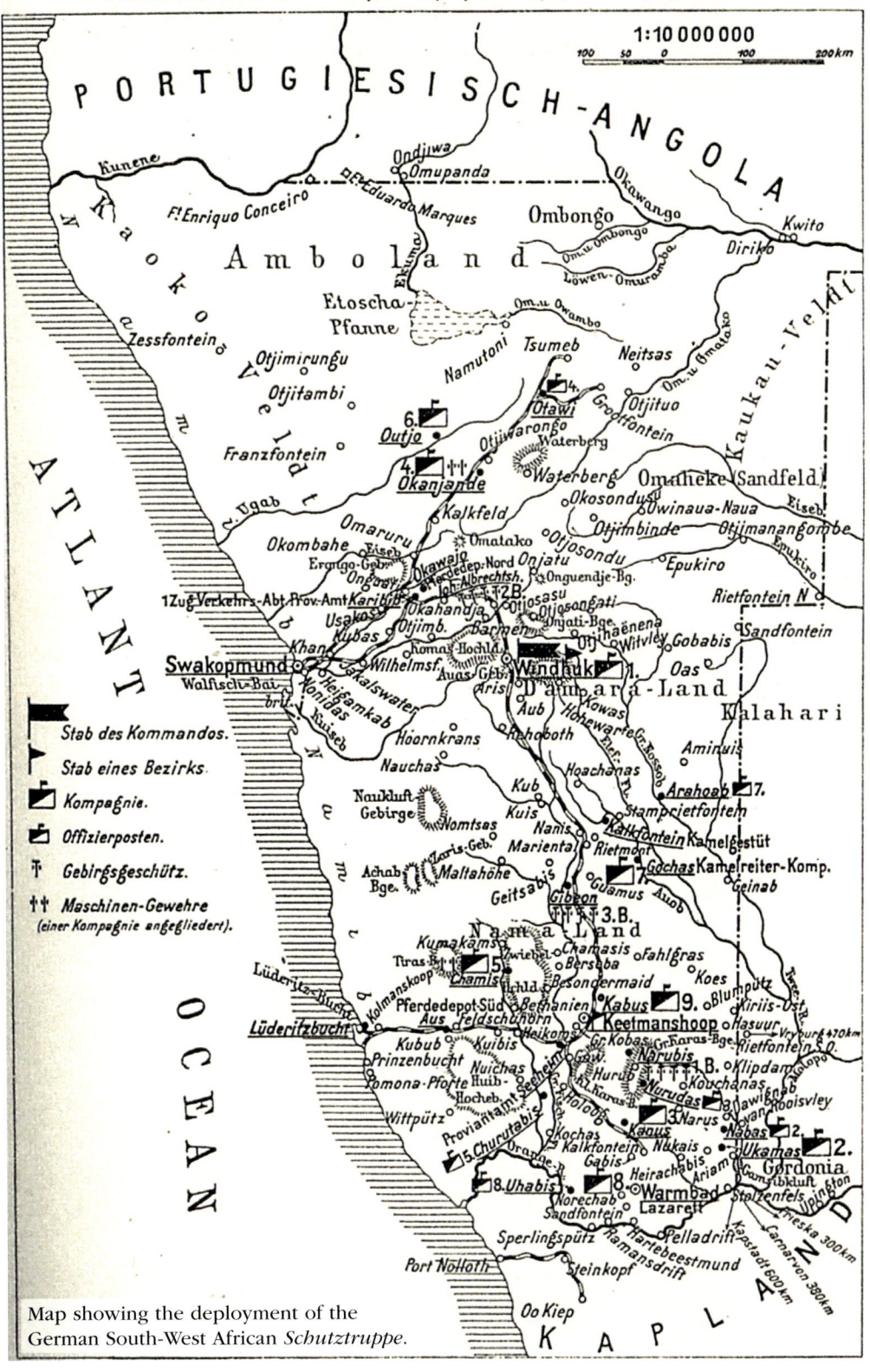

Map showing the deployment of the German South-West African *Schutztruppe*.

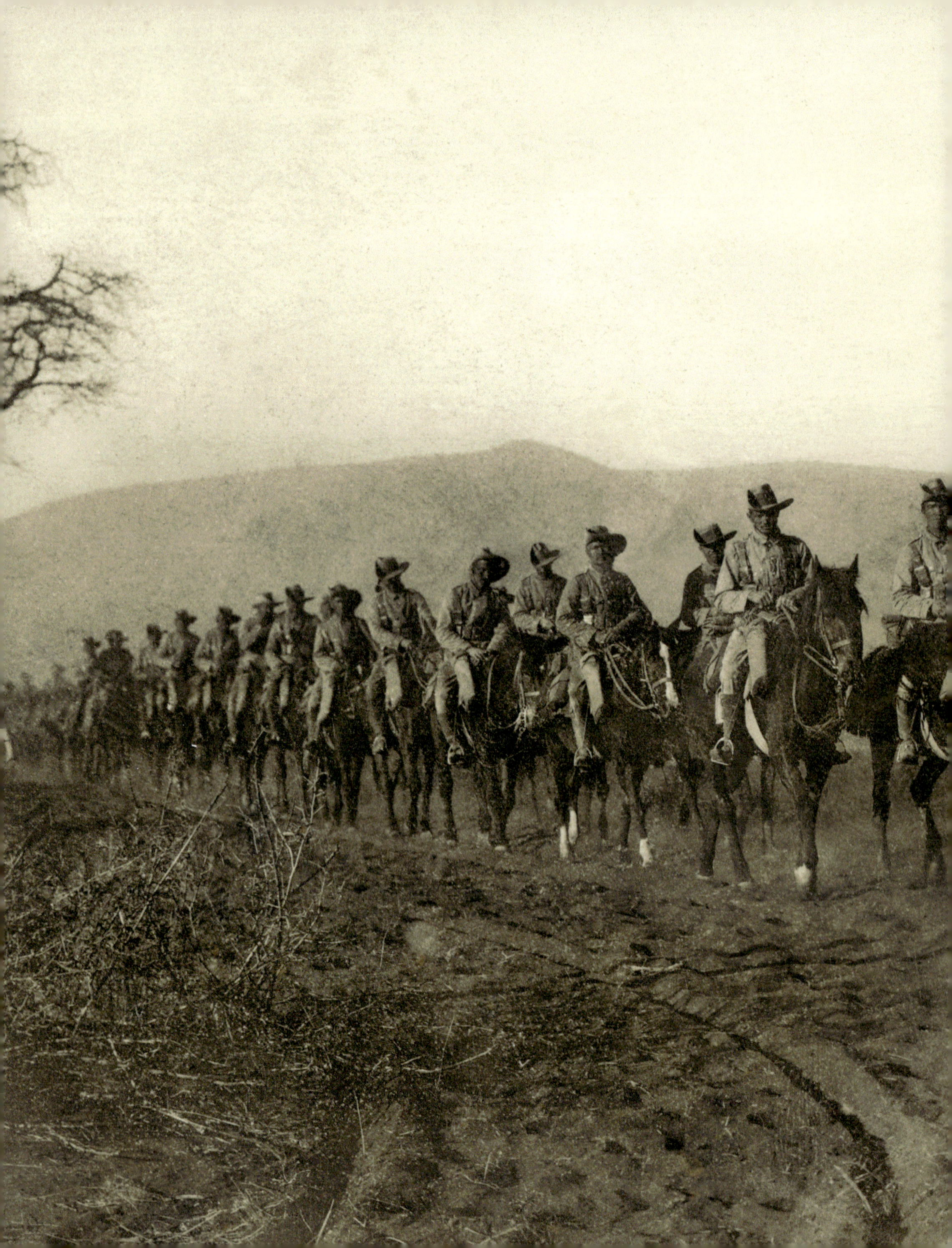

Mounted *Schutztruppe* Column on the march.

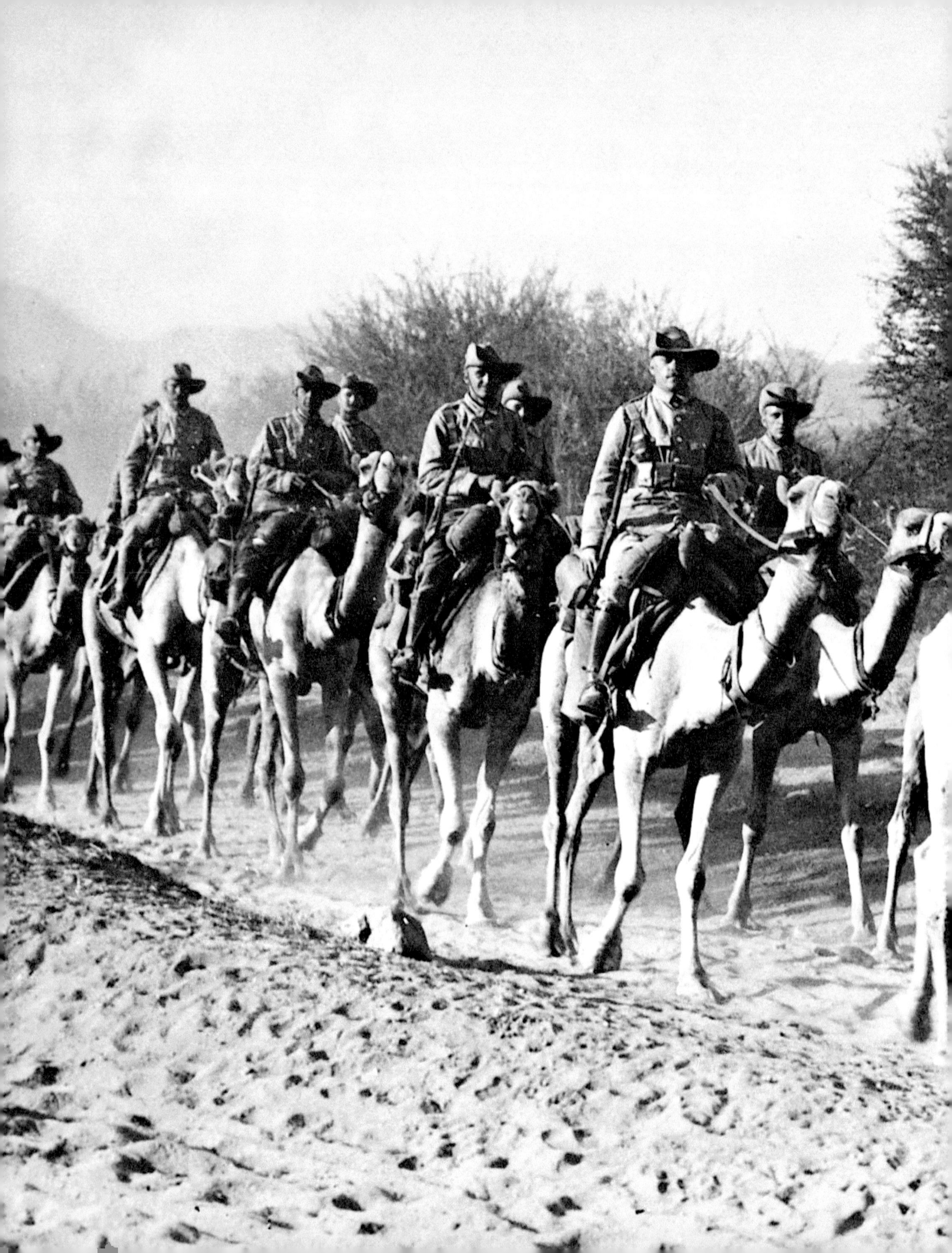

*Schutztruppe* 7th Camel Rider Company.
Photo: *Sueddeutsche Zeitung* / Alamy Stock Photo

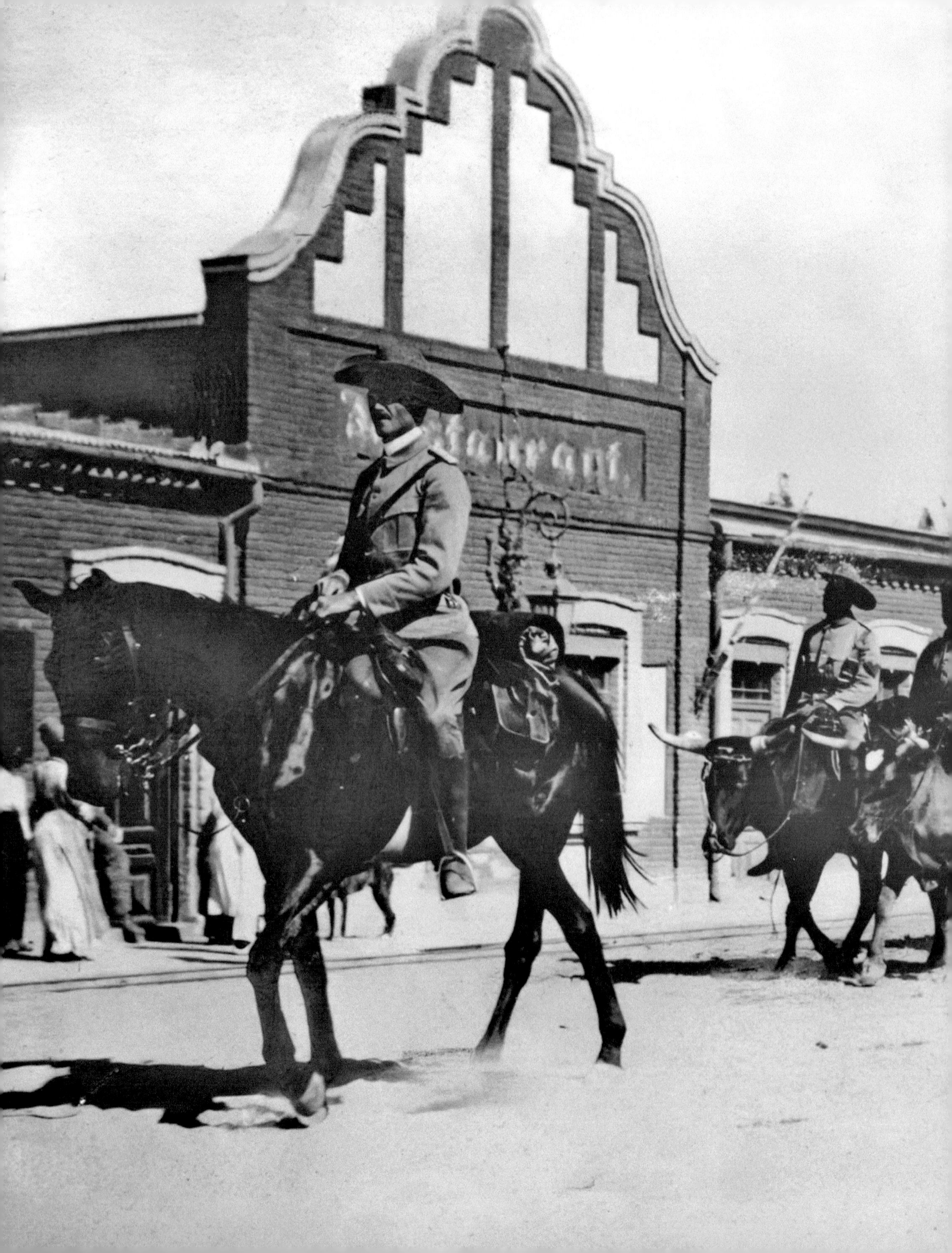

Oxen-mounted soldiers from the *Schutztruppe* Cameroon Company on the streets of Windhoek.

A *Schutztruppe* Column prepares to march.
In foreground: 7.2 cm *L/14 M98* Mountain Gun and Limber with four cases of Shells.

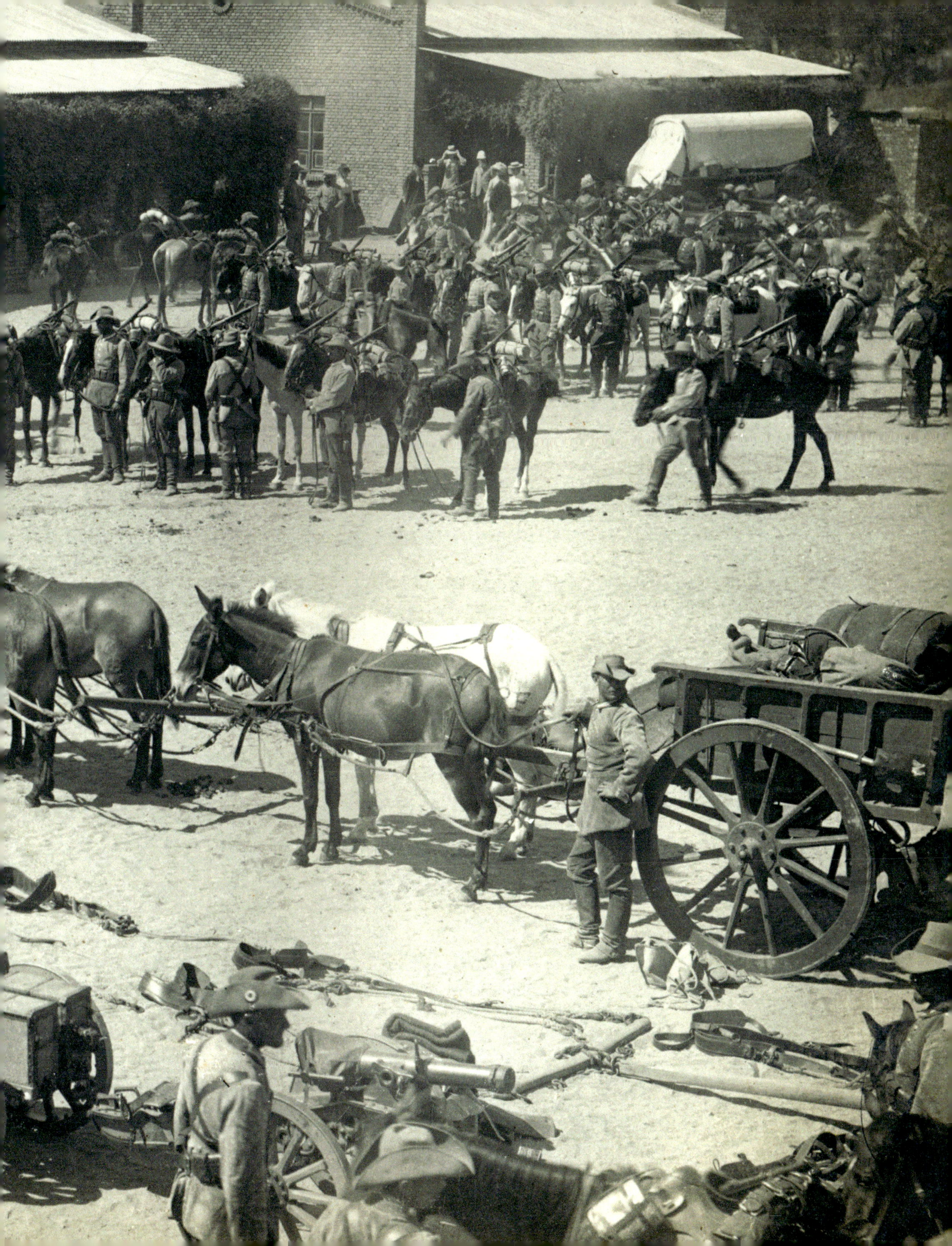

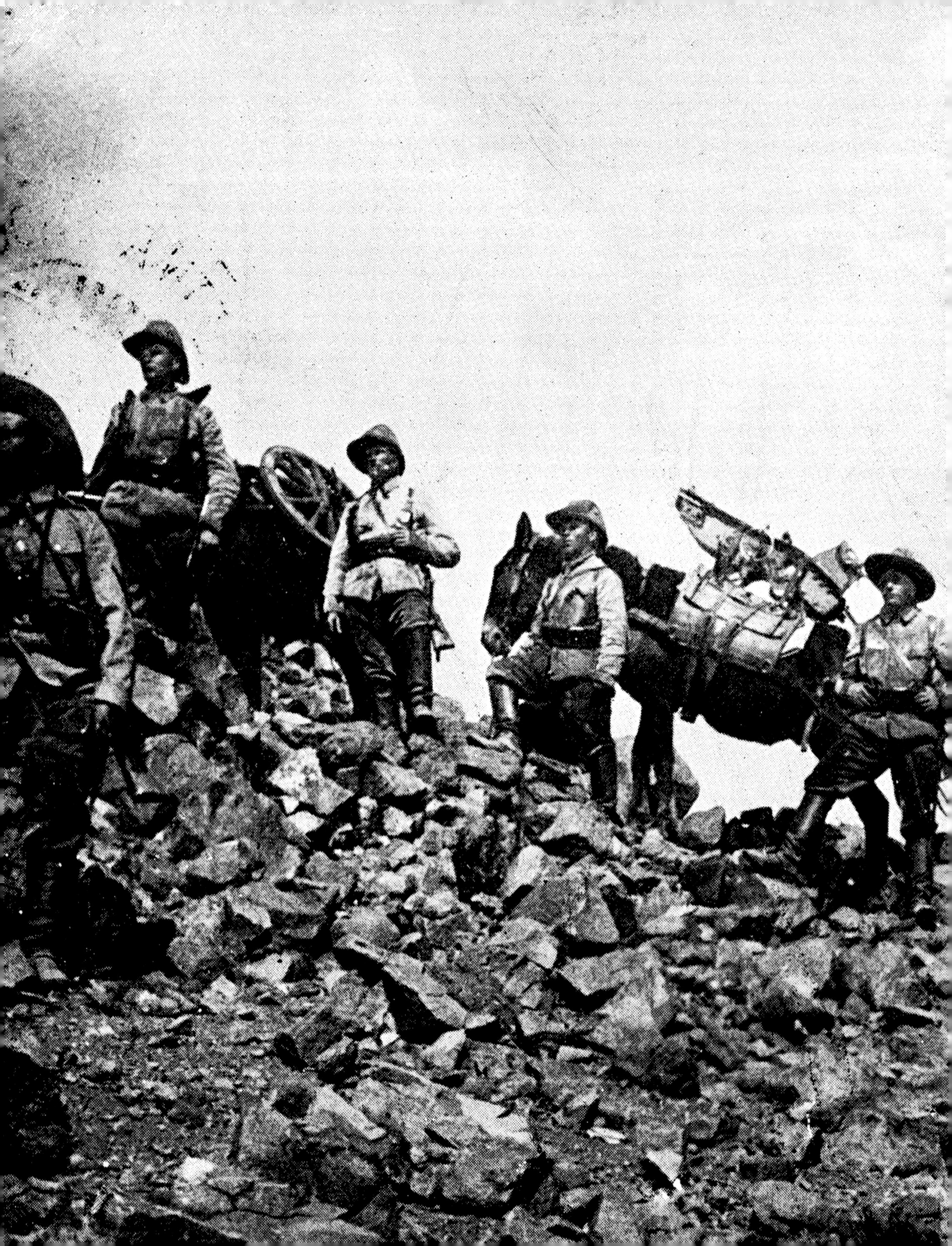

*Schutztruppe* Mountain Gun Crew operating in rocky terrain, 1911.

**Gruss aus Deutsch-Südwestafrika.**

Bedienung eines Gebirgs-Geschützes.

*Schutztruppe* Soldiers with 7.2 cm *L/14 M98* Mountain Gun.

Aus dem Lager der Gebirgsbatterie Deutsch-Süd-West-Afrika

Erhardt 7.5 cm Recoilless Mountain Gun in Towing Position.

Crew of 7.2 cm *L/14 M98* Mountain Gun at rest. Note open Ammunition Case and Shells next to Limber.

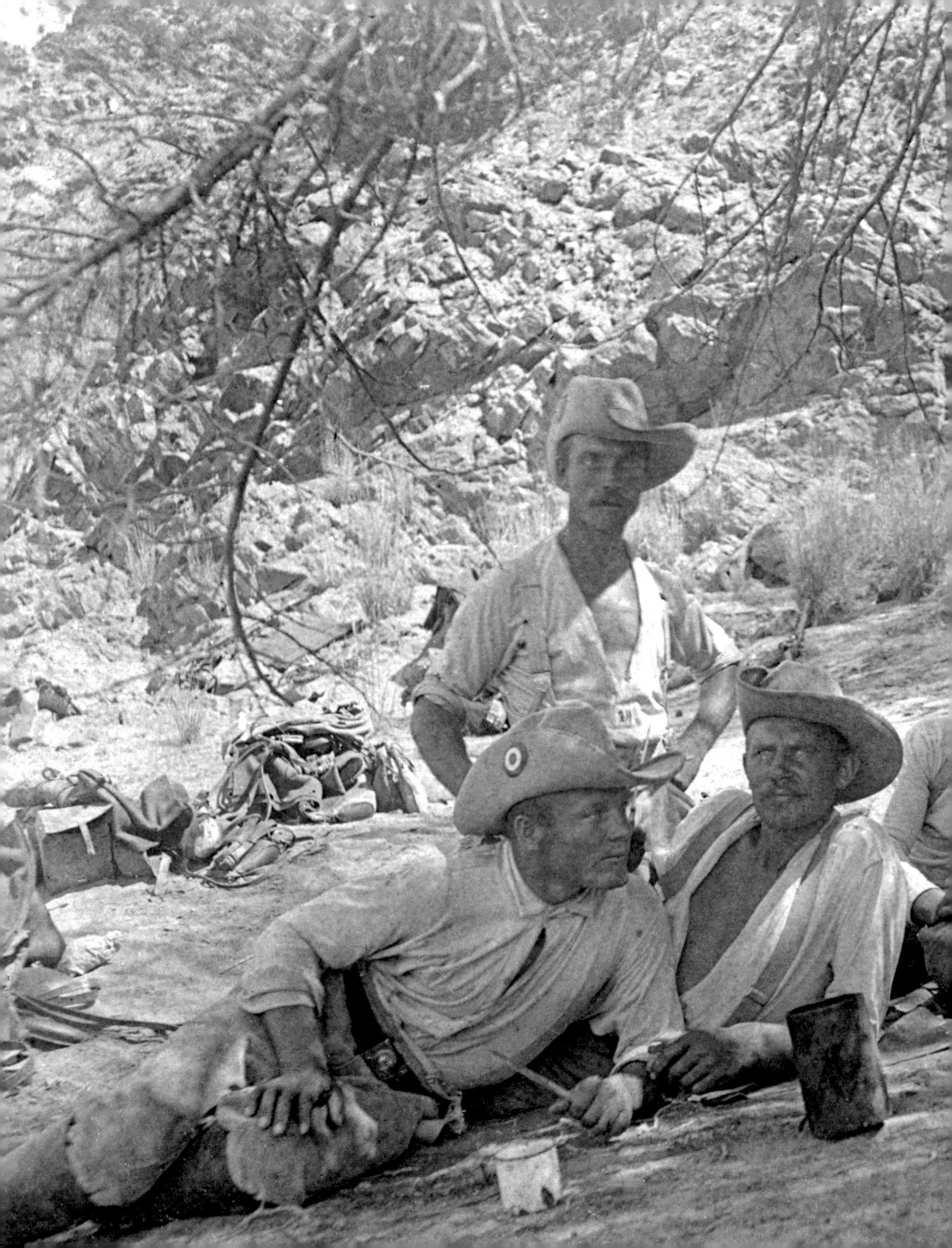

*Schutztruppe* Soldiers enjoying a meal in the shade.

Photo: *Süddeutsche Zeitung* / Alamy Stock Photo

Inset photos: Pausing in the field allows the dressing of minor wounds.
Below: *Schutztruppe* at Omaruru, 1904. Note German Marine Infantry in Tropical Helmets.

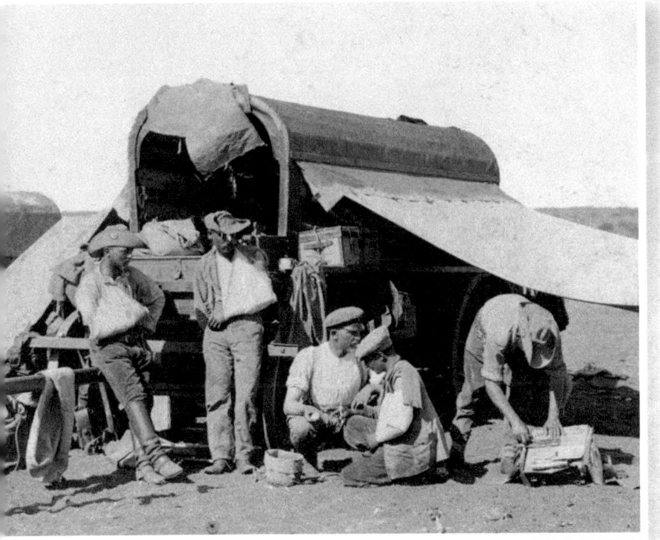

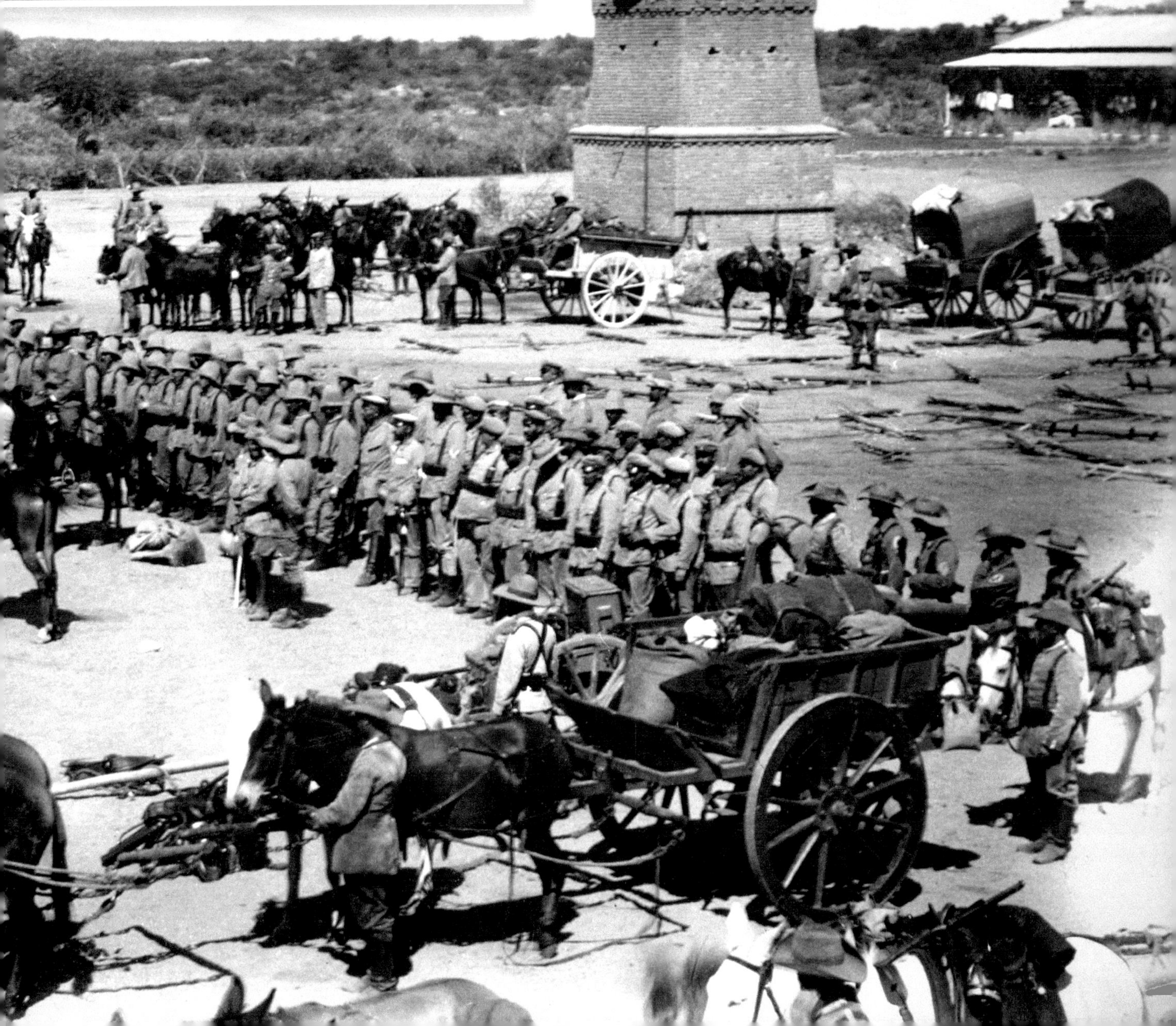

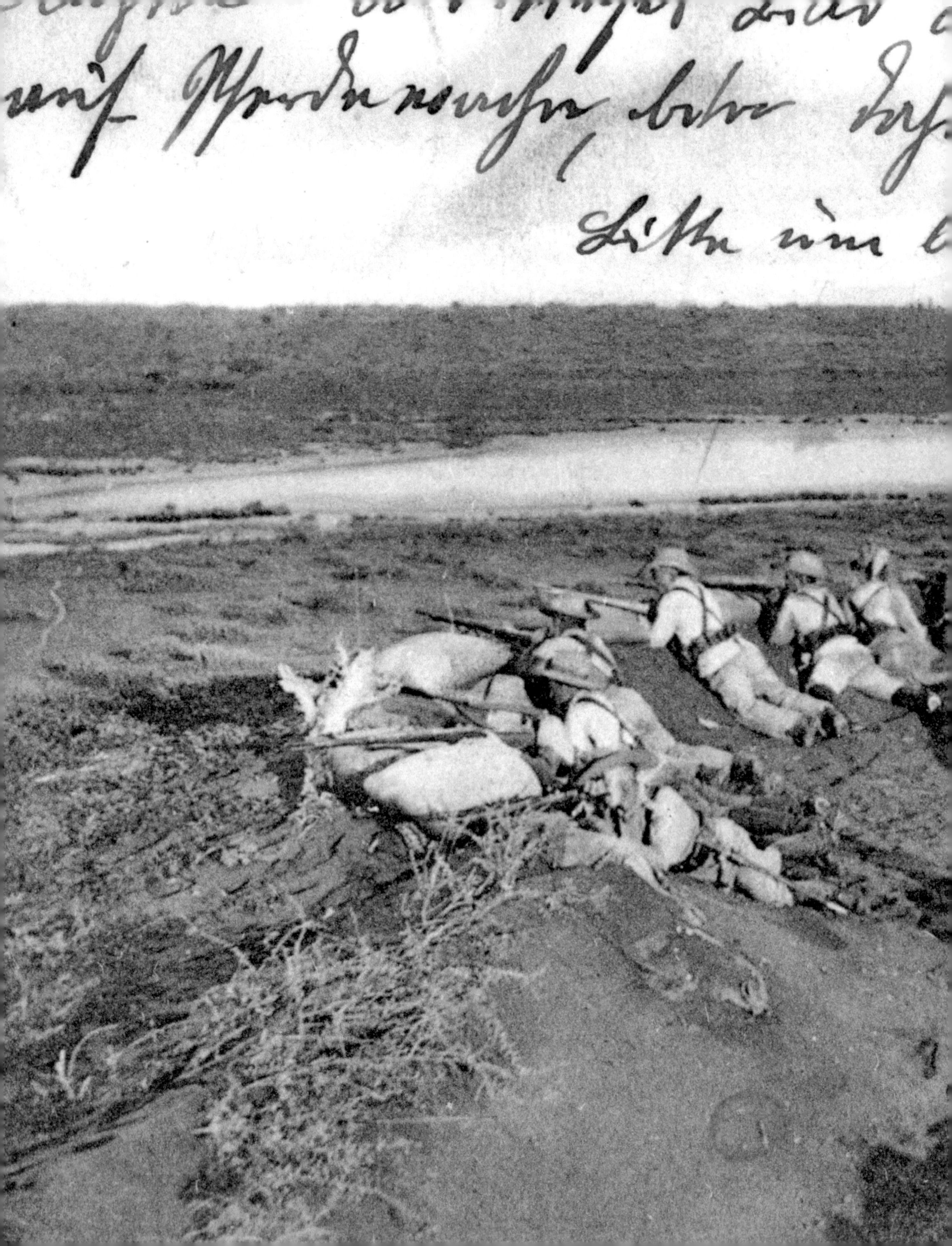

*Schutztruppe* Soldiers taking up a defensive position near a water hole.

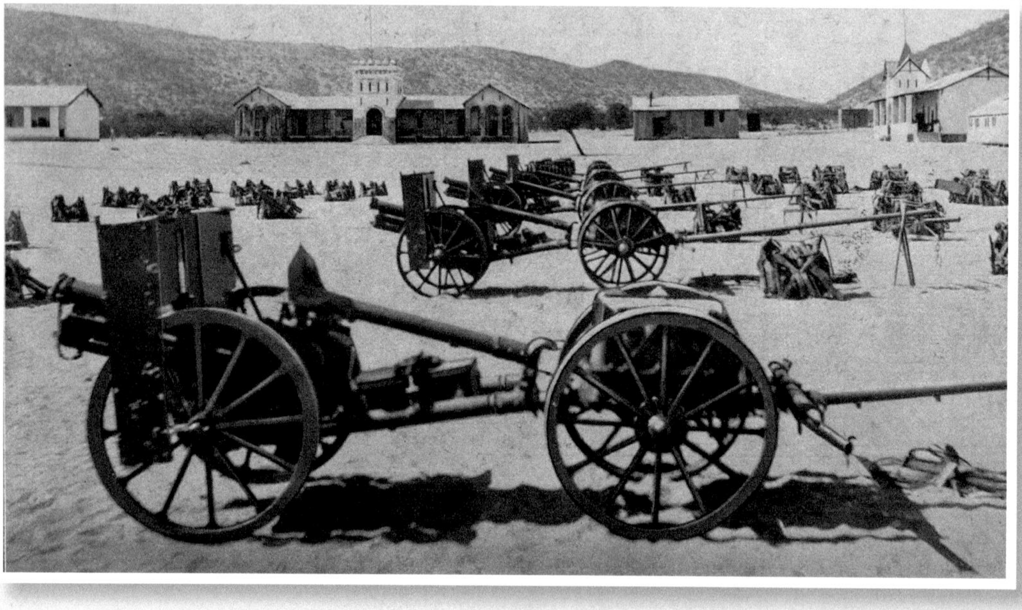

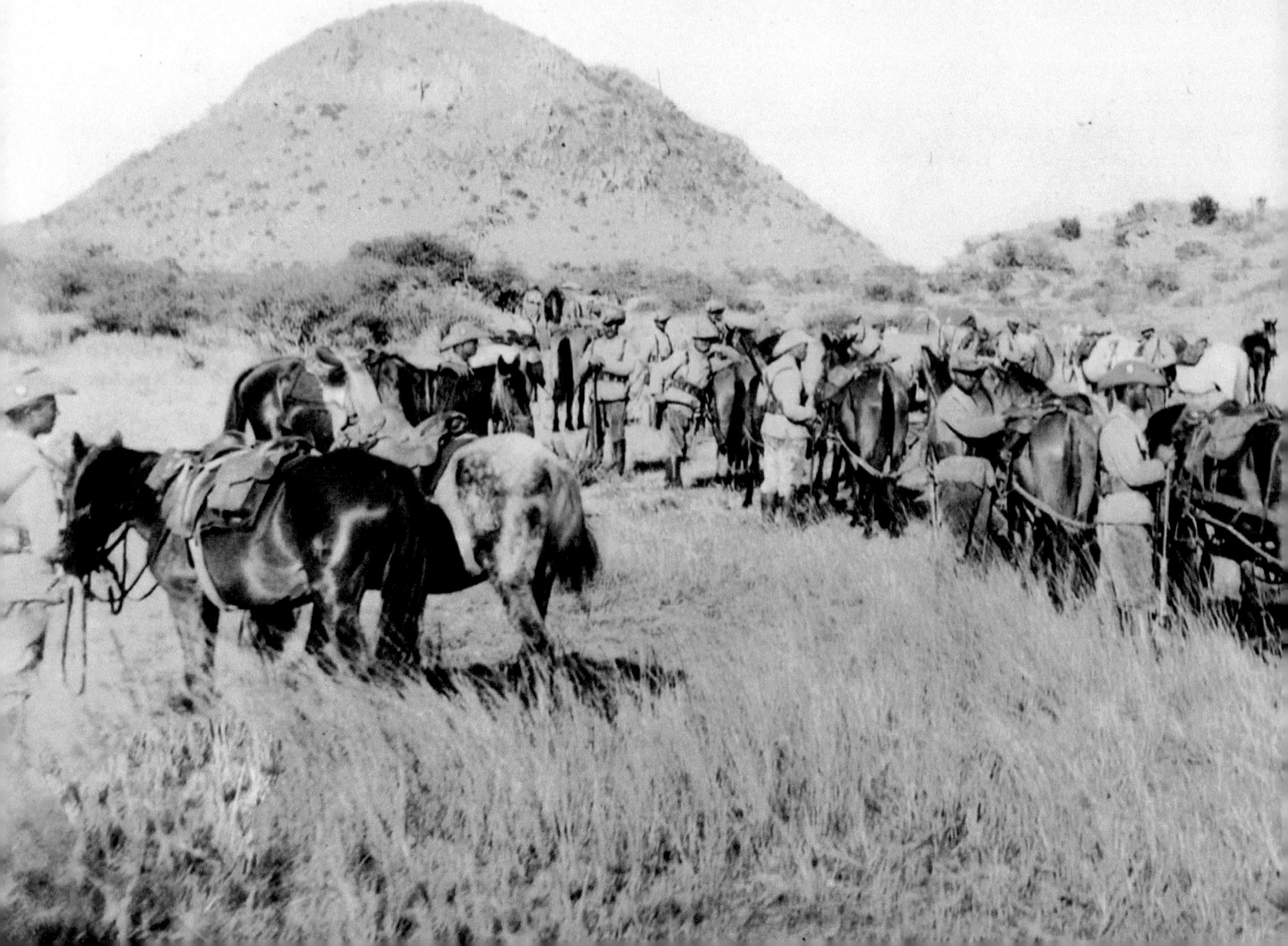

Main photo: *Schutztruppe* Gun Crew and Krupp *C96nA* Field Gun at rest in the foothills.
Inset: Erhardt 7.5 cm Recoilless Mountain Guns at Johann-Albrechts-Hoehe Station.

Large photo: Crew firing 7.85 cm *C73* Light Field Gun.

7.85 cm *C73* Light Field Gun and Crew.

Left: 7.2 cm *L/14 M98* Mountain Gun.
Right: 7.85 cm *C73* Light Field Gun and Limber.

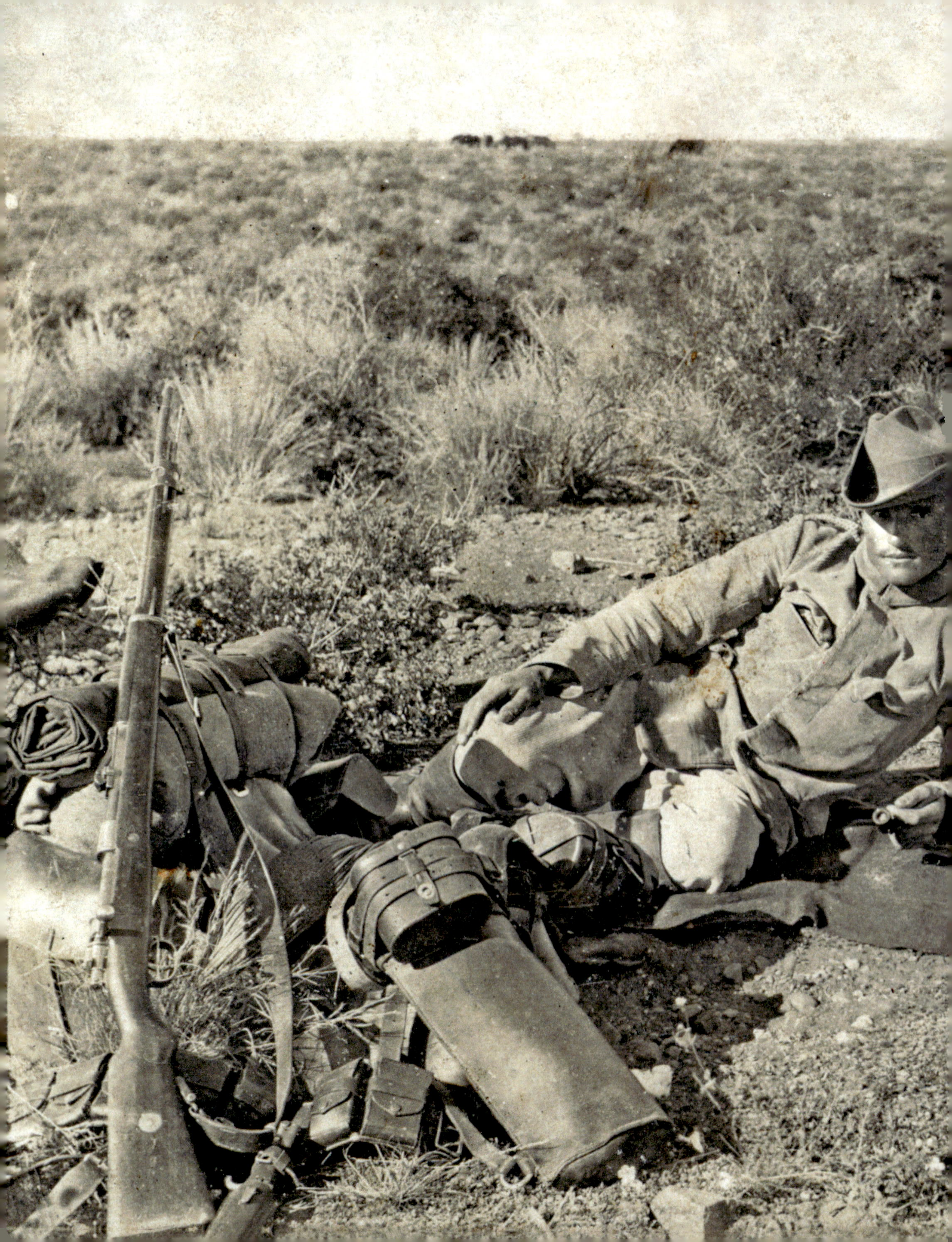

*Schutztruppe Sergeant* and *Unteroffizier* at rest, both wearing Corduroy *Litewka* Tunics and armed with *Gewehr M98* Rifles. Note Binocular Case strapped to Rifle Bucket and Compass on Blanket near Canteen.

Mounted *Landespolizei* on Parade.

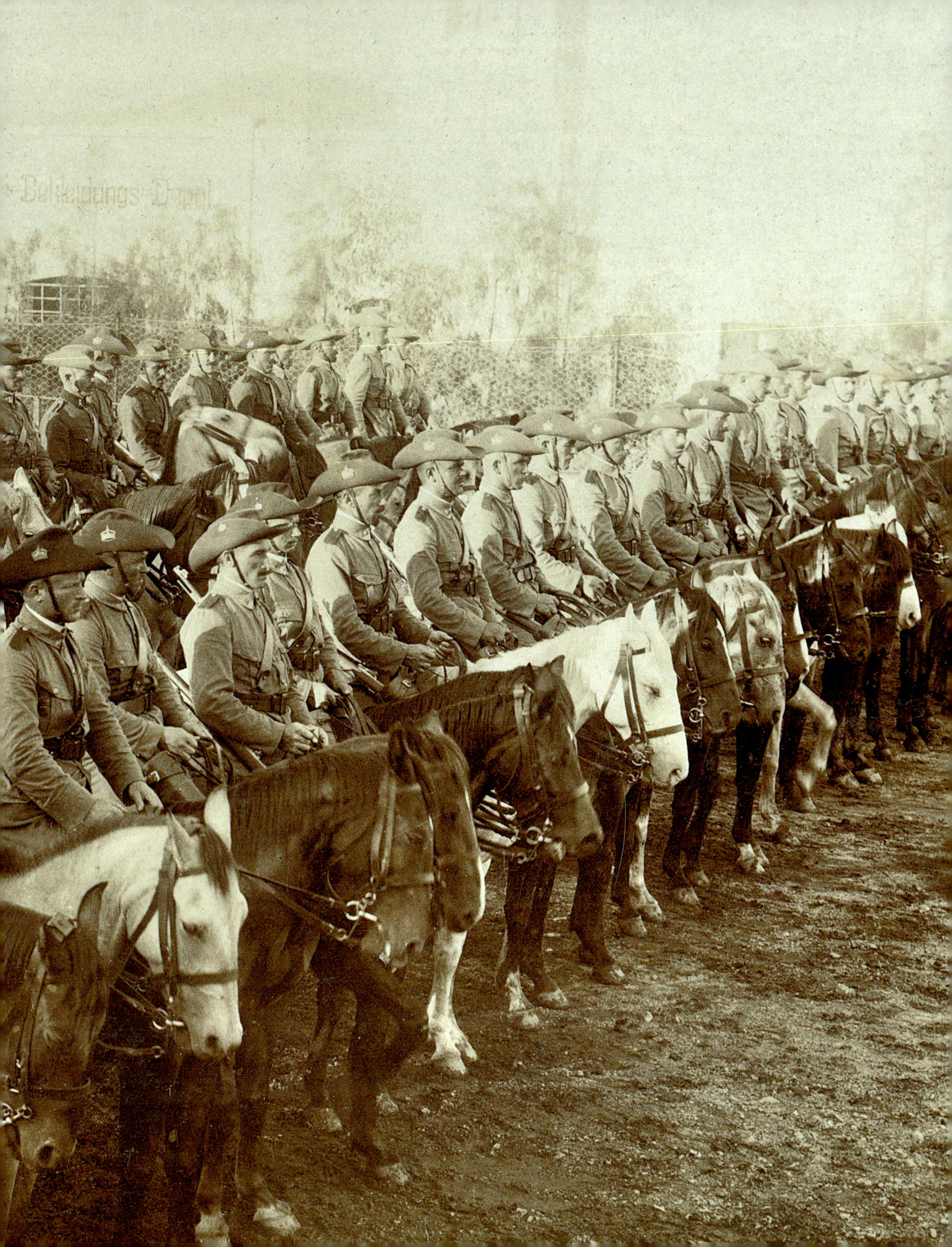

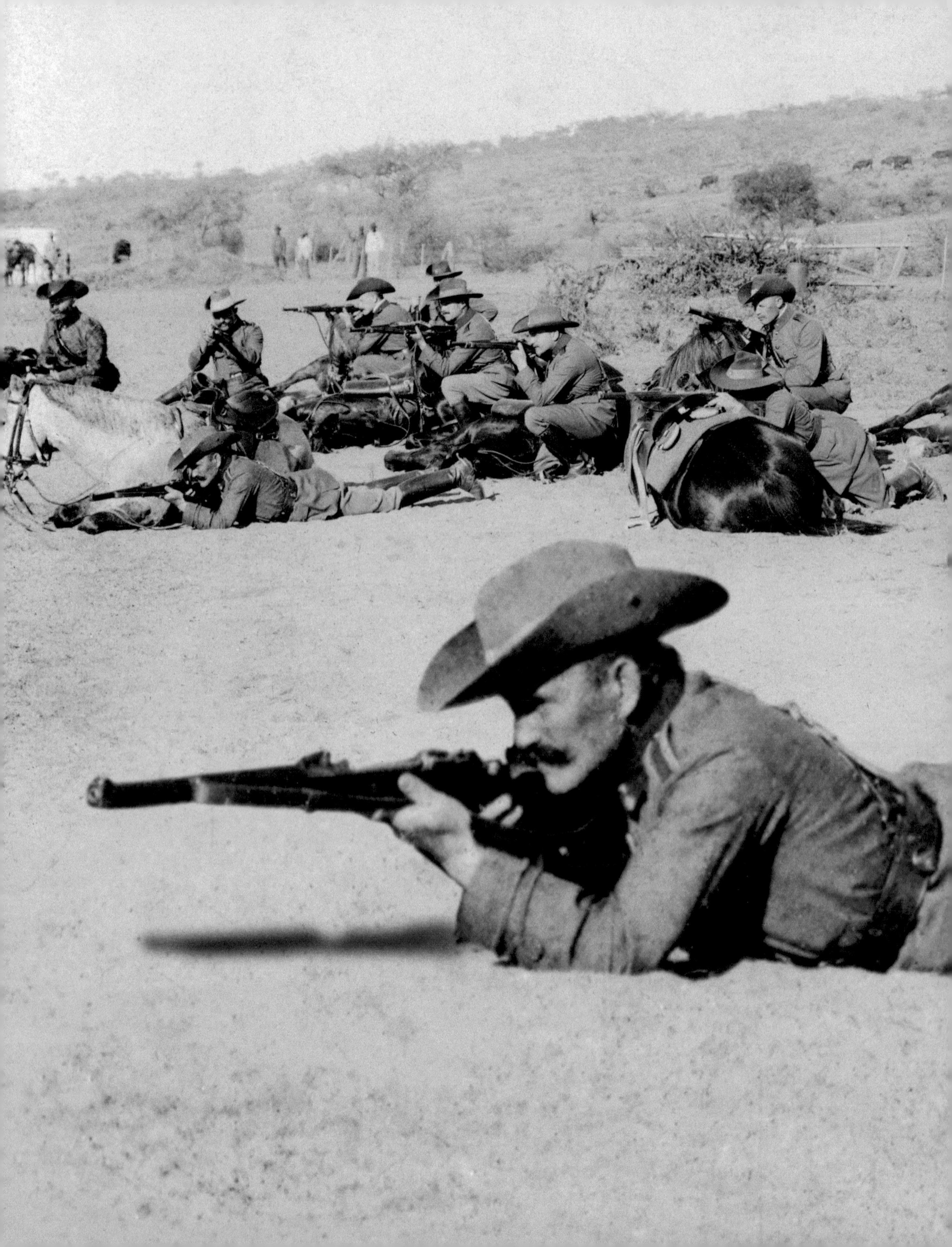

Composite photo of *Landespolizei* Troops practicing firing from behind their Horses.

*Schutztruppe* attend the unveiling of the *Reiterdenkmal* Statue in Windhoek, on the 53rd Birthday of Kaiser Wilhelm II, 27 January, 1912. Note the Swallows' Nest insignia on the Musicians' shoulders.

L to R: 86 mm flat figure and 90 mm *Schutztruppe* Buglers, painted by the author.

*Schutztruppe* of 3rd Field Company at Kanus celebrate their impending Home Leave.

HEIMAT

## *Schutztruppe* Anthem (To the tune of *'Stolz weht die Flagge'*)

I am a young blood rider in Imperial pay,
Wearing my hat above my ear, asking not for love nor gold,
With a brisk horse under me and a good rifle,
What else heaven brings me, does not heavily weigh on me.
    We serve you, dear fatherland in hot Africa;
    *Schutztruppe* we are called, and we are there to defend.

Although it burns as much as the sun on hot desert sand
With the brim around our wide hat, we go happily through the country.
Though our thirst torments us much, we are also tormented by hunger.
We all do not take it so hard, because all we think is that:
    We serve you, dear fatherland in hot Africa;
    *Schutztruppe* we are called, and we are there to defend.

We spend so many nights on foot and on horseback,
Defying sleep and keeping faith, it does not weigh us down.
For German brothers we do this gladly, and pride fills our breast,
For our Kaiser, our Lord, of this we are aware:
    We serve you, dear fatherland in hot Africa;
    *Schutztruppe* we are called, and we are there to defend.

We already fought a lot of bush and boulders,
With these black devils and their coats dyed red.
They pull out, we go behind as the sky turns blue,
And we rarely catch them, this they surely know:
    We serve you, dear fatherland in hot Africa;
    *Schutztruppe* we are called, and we are there to defend.

But alas, many comrades in their fullness of life,
There the rider fearless of death was gathered.
For our Kaiser, for the empire, he gave his blood,
On his face so deathly pale, there we lay his hat.
    We serve you, dear fatherland in hot Africa;
    *Schutztruppe* we are called, and we are there to defend.

And as our turn comes, too, we remain undaunted.
Through every battle life is refreshed.
We fight for German power, for German glory.
The distant homeland is faithfully remembered in joy and sorrow.
    We serve you, dear Fatherland, in hot Africa;
    *Schutztruppe* we are called, *Schutztruppe* on high! Hurrah!

## Schutztruppenlied

Ich bin ein junges Reiterblut in kaiserlichem Sold,
Trag auf dem Ohre keck den Hut, frag' nicht nach Lieb' und Gold.
Hab unter mir ein flottes Pferd und führ ein gut Gewehr,
Was sonst der Himmel mir beschert, das wiegt bei mir nicht schwer.
    Wir dienen dir, lieb Vaterland. im heißen Afrika;
    Schutztruppe werden wir genannt, zum Schutze sind wir da.

*Brennt noch so stark der Sonne Glut auf heißem Wüstensand,*
*Wir krempen um den breiten Hut und ziehn vergnügt durchs Land.*
*Quält uns der Durst auch noch so sehr, quält uns der Hunger schier,*
*Wir nehmens alle nicht so schwer, denn alle denken wir:*
    *Wir dienen dir, lieb Vaterland. im heißen Afrika;*
    *Schutztruppe werden wir genannt, zum Schutze sind wir da.*

*Und haben wir so manche Nacht zu Fuß und auch zu Pferd,*
*Dem Schlaf getrotzt und treu gewacht, es hat uns nicht beschwert.*
*Für deutsche Brüder tun wir's gern und Stolz füllt unsere Brust,*
*Für unsern Kaiser, unsern Herrn, denn wir sind uns bewußt:*
    *Wir dienen dir, lieb Vaterland. im heißen Afrika;*
    *Schutztruppe werden wir genannt, zum Schutze sind wir da.*

*Wir fochten schon so manchen Strauß in Busch und Felsgeröll,*
*Mit diesen schwarzen Teufeln aus und färbten rot ihr Fell.*
*Sie rissen aus, wir hinterdrein, soweit der Himmel blau,*
*Und holten wir sie selten ein, sie wissen doch genau:*
    *Wir dienen dir, lieb Vaterland. im heißen Afrika;*
    *Schutztruppe werden wir genannt, zum Schutze sind wir da.*

*Doch ach so mancher Kamerad in voller Lebenskraft,*
*Bei todeskühner Reitertat ward er dahin gerafft.*
*Für unsern Kaiser, für das Reich gab er dahin sein Blut,*
*Auf sein Gesicht so totenbleich, da legten wir den Hut.*
    *Wir dienen dir, lieb Vaterland. im heißen Afrika;*
    *Schutztruppe werden wir genannt, zum Schutze sind wir da.*

*Und kommt auch mal an uns die Reih', wir bleiben unverzagt,*
*In jedem Kampfe wird aufs Neu das Leben frisch gewagt.*
*Wir kämpfen ja für deutsche Macht, für deutsche Herrlichkeit.*
*Der fernen Heimat wird gedacht treulich in Freud und Leid.*
    *Wir dienen dir, lieb Vaterland, im heißen Afrika;*
    *Schutztruppe werden wir genannt, Schutztruppe, hoch, hurra!*

Deutsch-
Bezirke: 1 Kilimand
4 Pangani 5
Ssalam 8 Kilossa
11 Tabora 12
genburg 16
UGANDA
Mengo
Albert-Edward-S.
Victoria-
Nyansa
1180
KAWIRONDO
Virunga-B.
Bukoba
Kiwu-S.
1490
Karagwe
Schaschi
Katoto
Nassa
Ussukuma
Ruanda
Usige
Usindja
Muansa
12
BRITISCH
OST-AFRIKA
Kilimandjaro
Kwarango
Moschi
Ussiromba
Unyamwesi
Urundi
Uha
Manyara
Masai
13
Kirambo
Kilimani
Ussure
Kawele
Udjidji
Urambo
1240
Tabora
Masinde
Uwinsa
Unyanyembe
Kondoa
11
Tura
Kilima tinde
9
Tanganyika-See
Kawende
Ugalla
Igonda
Uyansi
10 Mpapwa
Nyambwa
Ugogo
Usagara
Karema
Ukonongo
Kilossa
8
Kirando
Mpimbue
Ukami
Kia
Urori
Uhehe
Marore
Khutu
Iringa
Fipa
Rikwa-See
14 Dwangiri
Mpueto
810
Niamkolo
Beja
Ulengula
Mahenge
15
Pambete
Rungue
Ugangi
Urungu
Konde
Wangemannshöh
Langenburg
Uemba
16
Marukutu
Moamba
Karonga
Muro-Alschinto
Livingstone-Geb.
Wangindo
18
Masasi
Newala
BRITISCH-
CENTRAL-AFRIKA
Bandawe
Nyassa-See
Rovuma
Mocambique
(Portugiesisch
30° Östl.L.v.Gr.
35

# Part II
# German East Africa
*(Deutsch Ostafrika)*

Map of German East Africa, 1906.

## Commanders of the *Schutztruppe*

1889 – 1891  *Reichskommissar* von Wissmann

1891  *Commandeur* von Zelewski

1892 – 1893  position vacant

1894 – 1895  *Oberst* Freiherr von Scheele

1896 – 1897  *Oberstleutnant* von Trotha

1898 – 1901  *Generalmajor* von Liebert

1901 – 1906  *Major* Graf von Götzen

1907 – 1914  *Major* von Schleinitz

1914 – 1918  *Oberstleutnant* (later General) von Lettow-Vorbeck

## Formation of the *Schutztruppe*

The German East African *Schutztruppe* emerged from the **Wissmanntruppe**, whose formation was prompted by the events of 1888 in East Africa. At the beginning of the year, a treaty had been signed between the Sultan of Zanzibar and the German East African Company, according to which the administration of the coast and the customs clearance in the name of the sultan should be transferred to the company. When the administration of the coastal customs levy began, the Arabs who had settled there centuries earlier began to revolt. They realised that their dominion over the indigenous population was over, and their vital interests were at stake.

With the help of natives of German Wituland, the German East African Company was able to suppress and take control of the entire coast, with the exception of Dar es Salaam and Bagamoyo.

At this point the German state intervened with an *AKO*. On 3 February 1889, then *Hauptmann* Hermann Wissmann was entrusted with the suppression of the Arab uprising. With the 'Law Concerning the Protection of German Interests and Combating the Slave Trade in East Africa' of 2 February 1889, Wissmann was appointed Imperial Commissioner (*Reichskommissar*) for East Africa. He had a close connection with the troops formed to fulfill his mission, which became known as the *Wissmantruppe*.

L to R: Seals of the German Colonial Office, *Schutztruppe* Command, and the German East-African *Schutztruppe*.

"I was the sole official of the empire and thus, with the means granted to me, responsible for the success (of the mission). As chancellor, I exercised due supervision of the German East African Company; all officers, officials, and the whole army were obliged only to me," wrote Wissmann in his notes.

Wissmann chose the African other ranks of his troop, and recruited the warlike **Sudanese** from Egypt as the core. Since the Sudanese themselves came from a humid tropical country, they tolerated the climate of East Africa

Governor Gustav Adolf von Goetzen and his Staff.

Photo: *Sueddeutsche Zeitung* / Alamy Stock Photo

**1889-1890**  Die Deutsche Schu

**Sudanese**  **Zulu**  **Sudanese**  **Swahili**

*Askari*  *Askari*  *Askari* Bugler  *Askari*

ztruppe für Ost-Afrika.  1891

These pages combine two different Moritz Ruhl lithographs. The four *Askari* soldiers on the left wear the khaki field uniforms from the *Wissmanntruppe* era (1889-1890). The three figures on the right depict a uniform approved 4. June 1891 for East African *Schutztruppe* German personnel and *Askari* uniforms of that era.

From a lithograph by **Moritz Ruhl, Leipzig.**

**Sergeant**

White Tropical Uniform

**Sudanese**

Khaki Uniform

**Sudanese Bugler**

Khaki Uniform

# Die Deutsche Schutztruppe

| **Lieutenant** | **Commander** | **Lieutenant** | **Medical Officer** |
| --- | --- | --- | --- |
| wearing the White Tropical Uniform | wearing the Blue Home Uniform as Gala Dress | wearing the Khaki Field Uniform | wearing the White Tropical Uniform |

für Ost-Afrika.  1891

These figures represent East African *Schutztruppe* uniforms for German personnel authorized 4. June 1891 and uniforms for *Askari* of that era. Figures have been moved to improve fit on page and German captions removed.

From a lithograph by **Moritz Ruhl, Leipzig.**

**Junior Paymaster**

wearing the
White Tropical Uniform

**Hospital Assistant**

wearing the
White Tropical Uniform

**Junior Gunsmith**

wearing the
White Tropical Uniform

# Die Deutsche Schutztruppe

From a lithograph by **Moritz Ruhl, Leipzig.** East African *Schutztruppe* uniforms for German personnel authorized 4. June 1891 and *Askari* uniforms of that era.

**Sergeant Major**

wearing the Blue Home Uniform

**Corporal**

wearing the Khaki Field Uniform

**Sergeant**

wearing the White Tropical Uniform

für Ost-Afrika.  1891

| **Askari** | **Askari** | **Askari** | **Askari** |
|---|---|---|---|
| NCO | | Bugler  2nd Company | 3rd Company |

Schutztruppe für

From a lithograph by **Moritz Ruhl, Leipzig**

East African *Schutztruppe* uniforms authorized for use in Africa 19. November 1896 and *Schutztruppe* home uniform authorized 11. March 1897. The illustrations mistakenly show cuff buttons on the white tunics.

**Paymaster**

wearing the
Grey Home Uniform

**Officer**

wearing the Grey
Home Uniform as Parade Dress

**Officer**

wearing the
White Tropical Uniform

# Ostafrika. 1896/1897

**Medical Officer**

wearing the
White Tropical Uniform

**Medical Officer**

wearing the
Greatcoat

**Hospital Assistant**

wearing the
White Tropical Uniform

**Sergeant**

wearing the
White Tropical Uniform

Schutztruppe für

From a lithograph by **Moritz Ruhl, Leipzig.**

East African *Schutztruppe* uniforms authorized for use in Africa 19. November 1896 and *Schutztruppe* home uniform authorized 11. March 1897. The illustrations mistakenly show cuff buttons on the khaki tunics.

**Sergeant Major**

wearing the
Grey Home Uniform

**Junior Paymaster**

wearing the
Khaki Uniform

**Ordnance NCO**

wearing the
Khaki Uniform

# Ostafrika. 1896/1897

**Senior Gunsmith**

wearing the
Grey Home Uniform

**Junior Gunsmith**

wearing the
Khaki Uniform

**Corporal**

wearing the
Greatcoat

**Corporal**

wearing the
Grey Home Uniform

*Schutztruppe* Column resting in a dry river bed.

well, although outbreaks of fever were not uncommon in the Sudanese companies. German officers at the time reported that the most notable feature of the Sudanese soldier was his extraordinary calm, which never left him, even in difficult situations.

At the same time, the Sudanese were very willing and zealous in their work – brave and, above all, reliable in the face of the enemy. On guard duty they were unsurpassed. The less favourable aspects of the Sudanese were their low marching ability and their apparent lack of a sense of direction.

**Zulus** were also hired, a quite different type of soldier material. These came from the southern part of the Portuguese colony of Mozambique, and were recruited in Inhambane. The Zulus had a quick mind, an astonishing

Forward March

*Leutnant* Graetz mounted on a Zebra with *Askari* ready to march.

sense of direction, an extraordinary marching ability, and, since they also coped well with the East African climate, made excellent recruiting material.

Indigenous soldiers (**Askari**) were represented only in small numbers. As a result of their knowledge of the country and the local language, they became interpreters and spies. They also formed the pioneer corps of the *Wissmanntruppe*.

To lead the African soldiers, Wissmann recruited a staff of officers and NCOs from the army. The supreme commander was *Reichskommissar* Wissmann, with the title of **Kommandant**. The unit was initially called the *Polizeitruppe* in East Africa, then the German African *Schutztruppe*, but it came to be known as the *Wissmanntruppe*.

The German officers were divided into the ranks of **Chef**, **Premier-Leutnant**, and **Seconde Leutnant.** Some of the officers were transferred directly from the army to the force of the *Reichskommissar*; others were former army officers serving as civil servants of the German East Africa Company.

The eight senior positions were given to the officers who had the longest record of colonial service, regardless of their seniority

Above and facing page: 5th Field Company on Parade for the Governor, September 1902.

in rank. One *Chef* was only an army *Landwehr Sergeant*. Upon entering service in the *Wissmanntruppe*, these officers retired from the army, although their later return to army ranks was assured.

The officers also included the doctors of the *Wissmanntruppe*: the **Senior Doctor** (*Chefarzt*) had the rank of **Chef** while the **Assistant Doctors** (*Assistenzärzte*) held the rank of **Premier** or **Seconde Leutnant**.

Ranked between officers and NCOs were the **Logistics Officers** (*Proviantmeister*), later graded as **Warrant Officers** (*Deckoffizier*) 1st and 2nd class. They were made up in part from lower officials of the German East African Company and other suitable Africans, and included **Junior Paymasters** (*Zahlmeister-Aspirant*). On occasion individual officials such as these saw action as platoon leaders. The German NCOs were divided into four ranks as in the imperial army: **Sergeant Major** (*Feldwebel*), **First Sergeant** (*Vizefeldwebel*), **Sergeant** (*Sergeant*) and **Corporal** (*Unteroffizier*). **Hospital Assistants** (*Lazarettgehilfe*) were counted as NCOs without a special designation.

Their strength was 25 officers (including two doctors and a paymaster), seven Warrant Officers, 56 NCOs including hospital assistants, six companies of Sudanese (totalling 600 men), a company of 100 Zulus, a locally recruited *Askari* unit of 80 men, 30 Sudanese in an artillery unit, and one section of 40 Somali boat crew.

Thus the *Wissmannt ruppe* totalled around 850 men. In addition there were 20 Turkish policemen. The German NCOs initially formed a sharpshooter unit consisting of 40 men, with the men later distributed among the companies.

The *Wissmanntruppe* was joined by the fleet of the *Reichskommissar* and the administration, which at first consisted of five and later three steamers. Of the steamers, each had a crew of a captain, a boatswain, a boatswain's mate, and two machinists with the rank of warrant officer or NCO and a few African crew. The administration of the fleet consisted of an official with the rank of *Chef* and eleven officers with the rank of *Leutnant*.

By an *AKO* of 22 March 1891, the German East Africa *Schutztruppe* was placed as an Imperial force alongside the army and navy. It was divided into ten companies. Of these, four were garrison companies guarding the five major coastal districts of Tanga, Bagamoyo, Dar es Salaam, Kilwa and Lindi; four were expedition companies, and two were replacement companies. Their strengths varied, based on the development of the colony and the needs of the force itself.

The colonial administration then passed from the *Reichskommissar* to the newly appointed colonial Governor and the fleet likewise became the Governor's fleet. At the time of the takeover, the *Reichskommissar*'s *Schutztruppe* had a strength of:

- 33 officers
- 5 doctors
- 16 warrant officers
- 64 NCOs
- 12 African officers
- 50 African NCOs
- 1,500 *Askari*

The now imperial *Schutztruppe* were assigned to the Naval Office, and their German personnel became soldiers under direct imperial service. *Reichskommissar* Wissmann retired to Germany for medical reasons.

## Uniforms of the *Schutztruppe*

The East African *Schutztruppe* consisted not only of officers, military officials and NCOs from the German army and navy, but also of African other ranks. There were also African NCOs and officers, although all German military personnel outranked Africans of any rank. Nor were German NCOs and men ever subordinate to African officers. Europeans and Africans made differing demands on tropical equipment. The uniforms were therefore different for German troops and for recruited Africans. This distinction was maintained until the First World War, and it was only the lack of supplies once the War began that finally led to Germans and Africans wearing the same uniform – as far as it is possible to describe their clothing as a "uniform" by the end of the war.

### German Officers and NCOs

Dark blue tunics and trousers were introduced as **Home Uniform** (*Heimatuniform*), **Parade Uniform** (*Paradeuniform*) and **Society Uniform** (*Gesellschaftsuniform*) upon the formation of the troop. The tunic had a low collar of dark blue cloth and a row of yellow (or in the case of Warrants Officers and officials, white) metal buttons down the front. On each side of the chest was a large patch pocket with flap with a similar pocket on both sides of the hips. The pocket flaps were fastened by small metal buttons. The tunic and trousers had no piping.

For garrison duty in Africa, the **White Uniform** was introduced in the same cut as the blue uniforms. Originally, they were made of white cotton and white flannel. The flannel garments were too hot and often did not last long in service. They were therefore soon abolished. Initially, officers also wore the white uniforms in the field. Later, white uniforms were worn only when in garrison. According to the regulations of 1890, they were used in **Service Dress** (*Dienstanzug*), for officers as well as for NCOs.

As well as the blue and white uniforms, for field use khaki-coloured uniforms were also worn. NCOs wore uniforms of tough grey cotton. Tunics made of reed linen and yellow khaki drill were worn as well but the linen proved unsuitable, as the color faded after

*General* von Lettow-Vorbeck,
54 mm figure painted by the author.

*Officer in Field Uniform and Askari standing at ease, 30 mm flat figures painted by the author.*

several washes. As a result, in the field and on marches only khaki drill uniforms were worn and prescribed by the regulations of 1890 for **Expedition Dress** (*Expeditionsanzug*).

Upon the formation of the *Wissmanntruppe*, rank insignia was introduced in black/silver/red braid. NCOs wore open chevrons on the left sleeve. The *Unteroffizier* had a single chevron, a *Sergeant* had two, while the *Feldwebel* had three. Officers and warrant officers wore the same braid on both sleeves of all clothing, in the style of the Imperial Navy. Warrant officers had a single bar of braid, lieutenants and doctors had two bars, a *Chef* and senior doctors had three and the *Kommandant* had four. Unfortunately, the dyes used in the black/silver/red braid were not color-fast, and after their first wash in Africa, all uniform parts were dyed to the elbows in the German colours. Thereafter, gold braid was added to the blue uniforms, and yellow to the white and khaki uniforms. The number of NCOs' chevrons and officers' bars remained unchanged for each rank. According to the regulation of 1890, the arm chevrons of the NCOs were now located on the left forearm.

The khaki drill **Field Tunic** was the same as worn by the South-West African *Schutztruppe* and described in that chapter, except that there were two sewn-in hooks in the rear tail skirts. Cornflower blue piping was worn, as previously described for the different ranks of the South-West African *Schutztruppe*, but this piping was not always present on the uniform. The shoulder straps were made of the silver braid, crossed with black and red silk threads, as already described for South-West Africa. The backing of the shoulder straps of the officers of the East African *Schutztruppe* was made of white cloth but otherwise the same as the South-West African *Schutztruppe*. The khaki **Field Trousers** were also the same as worn by the South-West African *Schutztruppe*.

The **White Tunic** (*Weisse Rock*) was for garrison and parade dress. It was made of white cotton fabric but otherwise had the same jacket-like shape, as well as other details as on the khaki field tunic. The **White Trousers** were also made of white cotton fabric, and had the

*Officer in Garrison Dress and Askari on the march, 30 mm flat figures painted by the author.*

Officer of the *Schutztruppe* in Field Dress, 54 mm figure painted by the author.

same cut as the khaki field trousers, but without the piping.

**Tropical Helmet** (*Tropenhelm*)

The old style pith helmet had a high top, a short front peak and long rear peak, and was made of cork covered with white cotton fabric. The front and rear peaks were edged in white and their underside lined in green. A mushroom-like ventilation cap covered with white cotton fabric was screwed on the top of the helmet. There was a 2.5 cm-wide white cotton ribbon around the helmet upon which **Officers** wore a 5 mm-wide silver cord known as the *Borgdatsch* and an embossed Imperial cockade. **Medical Officers** wore a golden cord. **Paymasters** had a silver cord and above the cockade wore a small silvered Imperial eagle. **Junior Paymasters** had the same insignia but without the silver cord. The **Ordnance Sergeant** had only the Imperial cockade. The **Senior Gunsmith** had a small bronzed Imperial eagle above the Imperial cockade. **Junior Gunsmiths** had a small brass Imperial eagle above the NCO cockade. **Senior NCOs** had the officers' more elaborate cockade, often made up of several parts, whereas the cockade for **Junior NCOs** and other ranks was stamped from one piece of metal. **Hospital Assistants** had the insignia of an NCO.

Both sunlight and rain badly affected the helmets, making them quite unpleasant. An attempt to overcome these problems came in the form of a waterproof khaki cover worn over the white helmet. Previously the need for camouflage in the field had often been taken care of by using khaki paint. The long neck shield of the tropical helmet was also found to be obstructive when shooting from a prone position.

Front and side views of an Officer's Tropical Helmet as used in German East Africa as well as Cameroon and Togo.

Uniform of a *DOA Schutztruppe* Officer, with *M08* Service Binoculars, *Gew M88* Rifle, and *M08* Pistol.

*Askari* Bugler, painted by W. Kuhnert.

A re-designed tropical helmet was shorter and covered with khaki drill, with a 2 cm-wide band of khaki drill which met at the back. At the top of the helmet was a 3.5 cm-wide hollow metal pommel, which was mushroom-shaped and covered with khaki drill, with three gaps in its edge for ventilation. This pommel was made of zinc brazed onto a brass bolt, which threaded into an open ring made of copper affixed to the top of the helmet. The helmet was lined with green drill cloth. The same green cloth was sewn in zig-zag fashion to hold a leather sweatband in place inside the helmet, and served to maintain a 5 mm gap through which air could flow up into the helmet and out through the top. The edging of the helmet was formed by a 2 cm-wide strip of khaki drill, fastened by machine stitching. On both sides of the inside of the helmet, brass hooks were located to attach a chin strap. The chin strap consisted of a 15 mm-wide cotton belt with brass slider buckles and triangular brass rings for attaching to the hooks. These buckles and rings were attached by copper rivets. Around the band of the helmet was a black/white/red cord with an Imperial cockade on the front. For junior NCOs this was made of cotton, whereas for senior NCOs it was silk. Officers wore a silver silk cord. A white helmet cover was worn on

*Leutnant* Graetz.

Signals Cadets practice using the Heliograph.

Heliograph Signal Station.

parade. It was made of white cotton in six parts with a 1.7 cm-wide band around the base, with a white cord to pull it tight. At the upper meeting point of the seams was an opening of about 2.5 cm diameter for the ventilation piece. The 1900 Bortfeldt helmet was protected by a fifteen-year patent. Until 12 September 1915, Ludwig Bortfeldt was the sole authorized supplier of this helmet to the Imperial Navy, *Schutztruppe*, and occasionally the Army.

## Südwester Hat

The grey felt hat was the same as described previously for officers of the South-West African *Schutztruppe*; however, in East Africa the hat band and trimming of the brim were made of white silk ribbon for officers and senior NCOs and white woollen ribbon for junior NCOs.

## Cap

The cap was made of grey corduroy, but with a hatband and piping around the brim in white silk ribbon for officers or white woollen ribbon for junior NCOs. In all other respects it was the same as for South-West Africa.

# German East Africa

*Askari* and German NCO in their Field Uniforms.

*Askari* making Camp and setting up their Tents.

*Askari* training in Camp.

## Shoes

The shoes were short leather lace-up boots made of brown leather. Strap-on gaiters were worn with the boots. Initially these were cloth, but later they were made of leather. Field grey puttees were also worn by some. For parade dress, black shoes were authorised. For ordinary garrison service, white canvas shoes were generally worn.

## Belt

The belt was a brown leather strap with a brass hook at the end. The brass buckle had a silvered badge with the Imperial crown.

Photo:
Fotoarchiv
Firma Helmut Weitze
Militärische Antiquitäten KG

*Schutztruppe* Officer's Belt.

## Equipment of the *Schutztruppe*

In Wissmann's time NCOs had the Prussian army **M1867** Canteen, while officers had a much larger flask, covered with felt and secured by a lock to keep the local servant from drinking from it. Later the newest army model was used. The army-pattern Bread Bag (*Brotbeutel*) was made of khaki-coloured cloth. A Pistol Holster, Rifle Bucket and Binoculars were also issued.

## Weapons of the *Schutztruppe*

Officers were armed with the **Reichsrevolver** and later the **Pistole M08** or other privately purchased pistols. Junior NCOs initially carried the **Gewehr M71/84** rifle or the **Karabiner M71** carbine. These were later replaced with the **Karabiner M98AZ**. These rifles were accompanied by the **S71/84** and later the **kS98** bayonet.

Die Uniformierung der deutsch-ostafrikanischen Schutztruppe.

1891 uniforms of the German East African *Schutztruppe*.

# Uniforms of the African Troops (*Askari*)

On 3 May 1889, the steamer *Martha* docked at Bagamoyo, bringing with it equipment made in both Germany and Cairo. This included uniforms, which were issued immediately. On 6 May the Sudanese Company headed for Dar es Salaam, wearing their new uniforms for the first time.

Die Uniformierung unserer afrikanischen Schutztruppen.

1896 uniforms of the German East African *Schutztruppe*.

Zulu *Askari* and *Schutztruppe* Officers with two Maxim Machine Guns, 1894.

Target Practice.

*Askari Betschausch*,
54 mm figure painted by the author.

*Askari* in Field Marching Order,
54 mm figure painted by the author.

Zulu Recruits after their issue of Uniforms, in front of the Fort (*Boma*) at Bagamoyo, ca. 1889.

Swahili *Askari,* the first East African Native Soldiers of the *Schutztruppe*, ca. 1889. The centre figure, Dunia, has had his hands chopped off by leader of the Arab revolt Bushiri bin Salim Al Harhi for cooperating with the Germans.

## Sudanese Companies

The Sudanese wore a brown khaki uniform similar to the army drill jackets, but with a stand-and-fall collar and shoulder straps. Leg wear consisted of half-length pants with blue puttees and natural-coloured lace-up leather short boots.

*Schutztruppe Askari,*
90 mm figure painted by the author.

The first Sudanese Recruits, ca. 1889.

*Askari* with Imperial German Flag,
86 mm flat figure painted by the author.

*Schausch* Imbarag Ali of the 12th Field Company in Mahenge.

Police Askari Fez with large Imperial eagle.

The first shipment of lace-up boots came from a clothing depot in Germany but were completely useless, as the shape of the African soldiers' feet did not match those of the Europeans for whom the shoes had been designed. The men therefore took off their shoes in combat and during long marches hung them over the bayonet.

There was also a red fez with a cloth turban wrapped around it. This was a piece of thin gauze in light gray or light yellow, approximately 3 metres long and 2 metres wide. It was first folded to a width of 12 cm, then wrapped over a special wooden frame and finally put on the red fez. The winding was such that a piece two hands-wide hung down the back as a neck veil. This headgear was heavy and warm and therefore not very practical. The different companies were distinguished by coloured cords, which were sewn onto the turban so that they formed a small loop over the middle of the forehead.

## Zulu Companies

The first equipment given the Zulus was a uniform jacket and trousers made of a lightweight and washable dark blue woollen fabric. The jacket was the same length as the khaki tunics of the Sudanese, but was of a looser cut. It had an oval neckline like a sailor's collar, though not cut so deeply at the front. The jacket was fastened at the front by four black horn buttons and had elbow-length sleeves.

The only military insignia was a narrow black/white/red band. This ran horizontally across the chest just below the top button, over

*Schausch* Tanganjika, *Askari* Medical NCO.

each shoulder, and again horizontally across the back. The trousers reached just below the knees. The Zulu troops marched barefoot but sometimes puttees were worn.

These blue uniforms did not prove practical. Starting around late 1889, the Zulus received uniforms similar to those of the Sudanese. A khaki uniform was issued with sleeves ending at the elbow and the pant legs just below the knee. The jacket had a naval cut rather than the simple stand-and-fall collar. The red fez was issued with company-coloured cord (1st Zulu Company had black, 2nd had white). Neither shoes nor puttees were issued.

### Swahili Askari

The Swahili *Askari* originally wore a white uniform similar to the Zulus' dark blue uniforms. The jacket had a similar band of black/white/red trim, but it was worn like a sailor's shirt, with white breeches. Later, the Swahili *Askari* received a high-necked white uniform jacket with long white pants. It was of the same cut as the Sudanese khaki jacket. A fez with a black tassel was the headgear, and no footwear were worn.

### Stations-Baharia

According to the regulations of 1890, the Stations-Baharia oarsmen were dressed in the white uniform originally worn by the Swahili *Askari*, with black/white/red trim. They wore as a special badge a blue anchor on the right sleeve. The leader of Baharia had a blue chevron on the left sleeve. Headdress was a straw hat.

## Die Grad-Abzeichen auf den Aermeln.

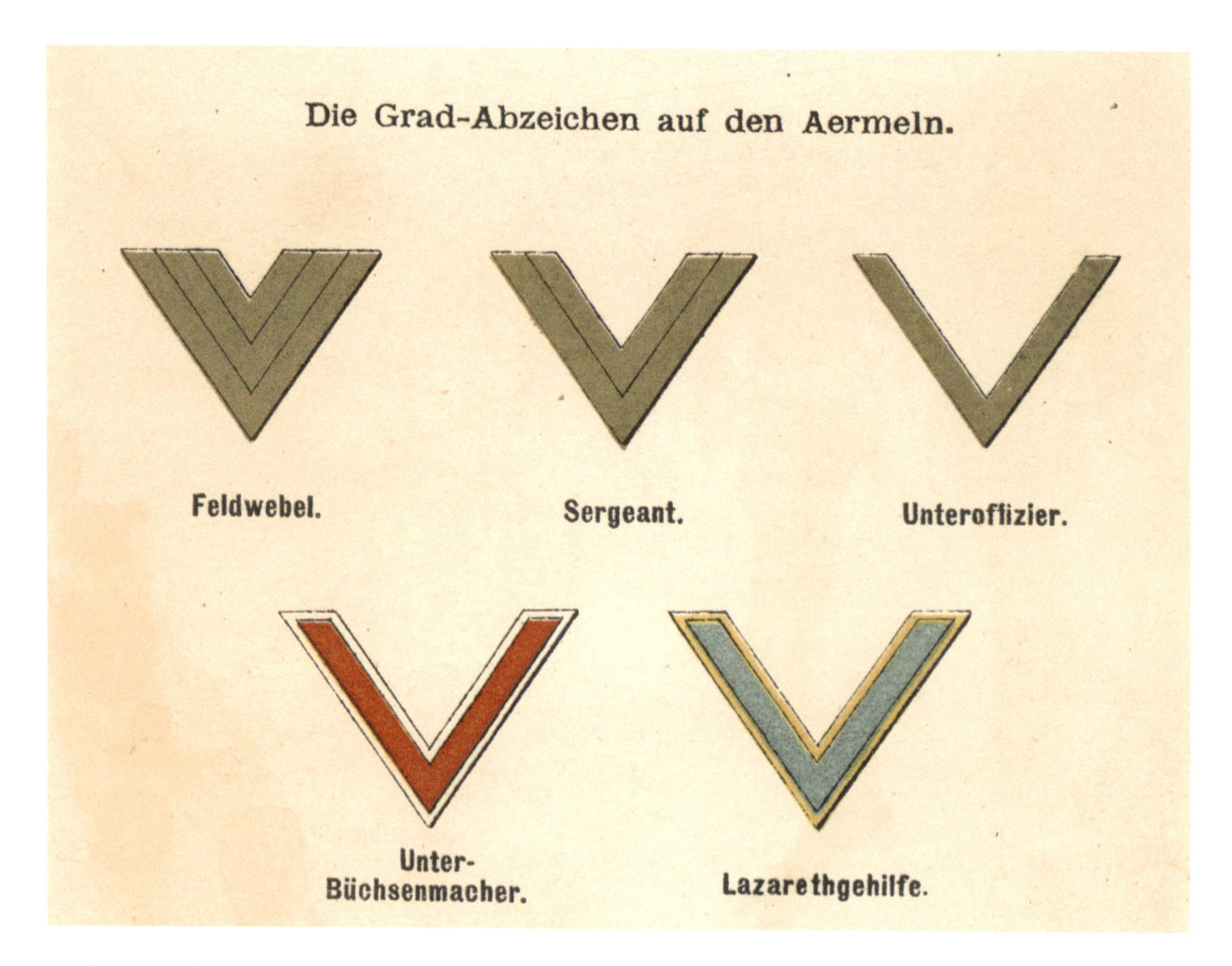

Feldwebel.     Sergeant.     Unteroffizier.

Unter-Büchsenmacher.     Lazarethgehilfe.

Rank Insignia for the German East African *Schutztruppe*.

## African Sailors of the *Reichskommissar's* Fleet

The **Sailors** of the *Reichskommissar's* fleet wore the same uniform as the Station Bacharia. On the straw hat they wore a tally band with their ship's name in gold (*Vesuv, Muenchen, Max*). They also wore the fez, and a blue anchor was worn on the chest. According to the regulation of 1890, the uniform of the sailors was left to the discretion of their captains.

## Standardized *Askari* Uniforms

Over the following few years the uniform was standardized after the pattern of the Sudanese uniform, and essentially remained so until the First World War.

The uniform of the *Askari* at this point consisted of a khaki, pocketless tunic with a stand-and-fall collar on which there was an eyelet with a hook to close it. Shoulder straps were of the same fabric, each with a small

Cover of accordion-style album of uniform illustrations, ca. 1891. Moritz Ruhl's multi-colour stone lithography plates are artistic and technical marvels. Often accompanied by detailed descriptive text from official uniform regulations, they are one of the best contemporary sources of information.

polished button. Down the front were five buttons made of bare metal. Khaki drill trousers without piping were worn along with brown leather lace-up ankle boots. On active duty a blue four-finger-wide bandage or puttee was wrapped from the top of the boot to the knee. In later years the puttees were grey.

**Tarbusch**

The *Tarbusch* now also became standard headgear. This was a peakless low cylinder made of straw or wicker, worn with a khaki cotton cover and a neck cloth that hung down to the shoulder blades. On average they were 12 cm tall, but deviations in height were quite normal. Initially the *Tarbusch* had a red twisted cord wrapped horizontally around the middle, looped in the centre.

This was replaced by a hollow brass embossed company number. The company number was later replaced by a silvered-metal flying

Imperial eagle. A red felt fez was worn when off-duty (not to be confused with the rolled red fez worn by native troops in Cameroon, which was much shorter and had a black wool tassel).

**Rank Insignia**

Rank insignia was worn on the left upper sleeve in the form of red chevrons, the upper ends of which formed a horizontal line. The ranks were:

- **Lance Corporal** – *Ombascha (Gefreiter)* :
  1 chevron

- **Corporal** – *Schausch (Unteroffizier)* :
  2 chevrons

- **Sergeant** – *Betschausch (Sergeant)* :
  3 chevrons

- **Sergeant Major** – *Sol* (*Feldwebel*) :
  4 chevrons

- **Officer** – *Effendi* :
  3 ranking stars on shoulder straps.

At first, there was also the rank of **Company Clerk** – *Bolukamin* – who were tasked with writing the private correspondence of the men. They ranked alongside the *Sol* or *Betschausch*. Most *Bolukamin* were soon sent back to Egypt because of their perceived laziness.

The *Effendi* and *Sol* did staff office duties at company level and served as intermediaries between the Germans and Sudanese. They also settled all disputes among the Sudanese. The difference between *Effendi* and *Sol* was that the former had a higher military education in Egyptian terms than the latter. Later the *Effendi* were phased out, and in 1913 there were only two *Effendi* in the *Schutztruppe*. These

Mule-mounted *Askari*.

*Askari* training in a defensive formation to fend off attacks in all directions.

African officers ranked the same as a German *Unteroffizier* or corporal. They carried a sabre without the officer's sword knot. Their uniform was in the cut of the German officers but without insignia. An African of *Premier-Leutnant* rank wore a single Egyptian star on his shoulder strap.

## Musicians

Musicians wore khaki drill swallows' nests with vertical trimming in red. The swallows' nests of the musician NCOs had 7 cm-long red fringe. The swallows' nests of senior musicians were made of red cloth with white trimming.

## Marksmanship Awards
(*Schützenabzeichen*)

The insignia for qualified marksmen consisted of black-white-red bands worn on both cuffs.

- 1st Award: one narrow band
- 2nd Award: two narrow bands
- 3rd Award: three narrow bands
- 4th Award: one broad band
- 5th Award: one broad band & one narrow band.

And so on upwards.

*Askari* of the early 1890s.

*Askari* Field Exercise with Machine Guns.

# German East Africa

Home Uniform of a *Leutnant*
in the German East African *Schutztruppe*.

German East African *Schutztruppe*
NCO's Cap.

German East African *Schutztruppe*
Officer's *Südwester* Hat.

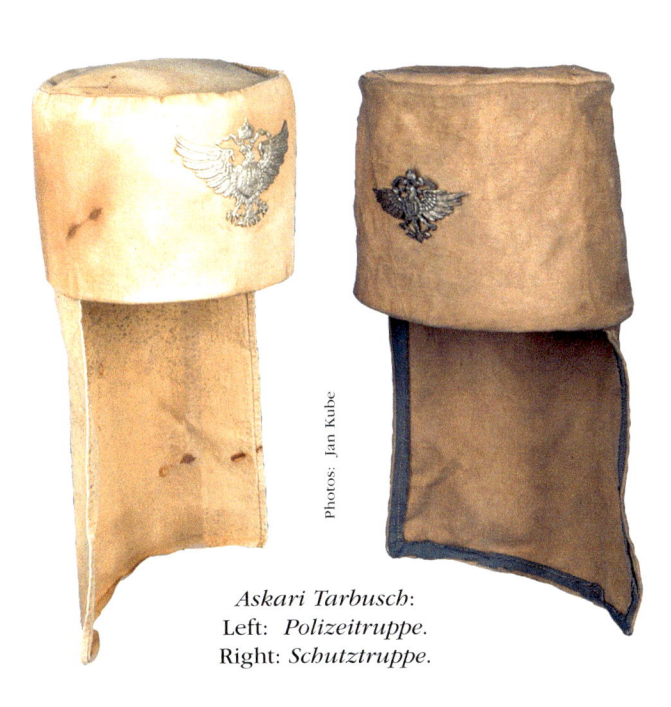

*Askari Tarbusch* of the
15th Field Company
with brass Company
number.

*Askari Tarbusch*:
Left: *Polizeitruppe*.
Right: *Schutztruppe*.

*Askari Tarbusch* Eagle.

Photos: Jan Kube

Photos [*]: Fotoarchiv Firma Helmut Weitze Militärische Antiquitäten KG

*Wissmanntruppe* in Bagamoyo, 1889.

Photos: Fotoarchiv Firma Helmut Weitze Militärische Antiquitäten KG

*Schutztruppe M1898* Senior NCO's Sword.

German East Africa 143

A *Schutztruppe* Officer's Quarters. Note the Tropical Helmet, *Jägerbüchse M71* Rifle and Sword with an Officer's Sword Knot.

*Feldwebel* Birkner wearing the *Schutztruppe* Home Uniform.

Uniform of an *Askari* in the 7th Field Company, with the *Jägerbüchse M71* Rifle.

## Artillery

Gun layers wore a red cloth flaming grenade on a white oval cloth backing on the left upper sleeve.

## Signallers

Signal personnel had a white patch with two crossed red flags on the upper left sleeve.

## Porters (*Träger*)

The *Schutztruppe* company porters were uniformed like the *Askari*, except that they had a single cartridge belt and the *Karabiner M71* carbine. Their red fez showed the number of their company on the front.

## Equipment of the *Askari*

The equipment included a brown leather belt with a smooth brass clasp. Normally two brown cartridge pouches were carried on the belt, with a third pouch carried on the back of the belt when on expedition. This third pouch was later discontinued, as experience had shown that in addition to the front pouches, soldiers often carried a self-made bandolier. Later this was also banned, as too many cartridges were lost.

A Y-shaped arrangement of leather straps served as a carrying frame. It was comprised

Artillery Depot in Dar es Salaam, including a 3.7 cm Quick-Firing Gun in the foreground and several *C73* Field Guns.

of two shoulder straps which hooked onto each of the cartridge pouches in front, and a narrower third strap which reached down the back and hooked under the belt. Hanging from the belt on the wearer's left side, behind the cartridge pouch, was a brown bayonet frog, in which the *S71/84* bayonet was sheathed in a leather scabbard.

The Canteen was attached to a strap worn over the left shoulder. A large cup turned upside-down fit over the body of the canteen, and a leather strap held them together. Later the canteen was carried without a cup. The Bread Bag was also of the pattern carried by the army.

There was also a Knapsack made of light fabric, similar to those used by the British army. The contents of the knapsack included a full uniform, sandals to wear in place of boots, plate, spoon, bags of salt, a sewing kit, rifle-cleaning utensils, and a small, sealed jar of ammonia for use against venomous arrows and bites from snakes, scorpions, centipedes etc..

Instead of a coat, the *Askari* had a light woollen blanket just big enough for the man to wrap in. This was rolled up inside a lightweight waterproof sheet and tied into a long bundle, which was positioned either over the knapsack or worn across the chest. This waterproof sheet also served as a sleeping mat.

Each man also had a tent – a 2 metre-long and 1.6 metre-wide piece of densely woven, lightweight cotton, with eyelets on its edges for tent pegs. It was stored under the flap of the knapsack. With the aid of two 60 cm-high supports and a 2 metre-long tent pole (which did not need to be dead straight), every man was able to construct his own tent within a few minutes.

Column of *Askari* and Porters on the march.

## Weapons of the *Askari*

Armament consisted of the *Gewehr M1871* infantry rifle, *Jägerbüchse M1871* light infantry rifle or *Gewehr M1871/84* infantry rifle, and later the *Gewehr M1898* infantry rifle. These were accompanied by the *S71/84* bayonet and later the short *kS98* bayonet.

The *Jägerbüchse M71* served to arm the African troops in the German colonies and was still in use during the First World War. When war broke out, only the *Askari* of the 1st, 4th, 8th, 10th and 13th field companies were equipped with the *Gewehr M98*.

In a modified form, some *M71* and *M71/84* rifles found a special use in combat during the World War. They were re-bored to a calibre of 12 mm, and Lentz incendiary rounds were fired from them. These were used to shoot down tethered balloons.

On the subject of the armament of the African soldiers in the colonies, another innovative measure is mentioned here so as not to be forgotten in the history of German weapons.

Around 1909 – 1910, attempts were made to alter the *M71* cartridge for the colonies with a low-smoke nitro powder. The reasons for this were twofold: the authorities wanted to keep the large-calibre 25 gram heavy bullets because of their reliable effect, but they also wanted to avoid the smoke from the black powder used in them. In the bush and tall grass, the thick clouds of smoke were particularly unhelpful, as they prevented accurate targeting of the enemy after the first few shots had been fired.

The hope was to keep the weapon, cartridge case and projectile, but to exchange the black powder charge for a charge of nitro powder. Experiments were made with a slightly porous powder paste, but the trials failed completely and provided evidence of the close connection between an effective weapon and its ammunition.

Of course one cannot arbitrarily change one element of an overall formula without throwing the other elements out of balance. The properties of the original black powder gave the projectile the required compression and guidance, even when fired at maximum range. The test results of the new ammunition yielded shameful results: when fired at a range of 300 metres, 10 – 20 % of shots fell limp in the sand at a distance of only 50 metres. Finally, the authorities were forced to give up all such attempts.

The Governor of German East Africa then advocated the gradual re-arming of the *Askari* with modern multi-round rifles. However, the fear of multi-loaders in the hands of African soldiers had already led to the removal of the *M71/84*'s multi-loading function.

The aforementioned five field companies with the *M98* rifles may then be considered as marking the beginning of these modernization measures.

## Marching on Campaign
(*Kriegsmärsche*)

Marching a company of soldiers and a support column in single file, in view of the many obstacles on the way, often caused gaps to form in the ranks. Thus a march into battle took a disproportionately long time relative to

*Askari* in Rifle Training.

the strength of the unit. These gaps within the marching column often meant that a deployment to the flanks and the concentration of troops in case of ambush became more difficult. A troop of 1,000 men traveling as lightly as possible could make about 5 km per hour on a good march, and required at least 45 minutes to start the march forward. This could be an eternity when facing an agile opponent.

The length of a marching column did not always increase evenly in proportion to the strength of the unit. This was caused by the aforementioned gaps in the ranks, which increased exponentially with the strength of the marching unit. For example, 100 men would stretch over 250 metres, 400 over 1,500 metres and 600 men over 2,500 metres.

A column of native porters needed even more marching room than the troop column. Yet even a unit with no artillery could not function without porters. Europeans of the *Schutztruppe* needed a certain amount of comfort to remain fit for action. A tent, a bed with woolen blankets, uniforms and a change of linen, cooking utensils, and some food required two to four porter loads for each European soldier.

The question of ammunition also came up for consideration. One could load a soldier armed with the *M71* rifle with no more than 100 rounds. This was precious little ammunition in the event of a bush fight in which the control of fire did not remain in the hands of the officers. *Askari* were more inclined towards ill-considered rapid fire than European soldiers, so extra ammunition was always necessary. A porter could carry 400 to 500 rounds, which was sufficient ammunition for only five men. It was difficult to find middle ground between the amount of ammunition required and the need to reduce the length of the porter column which hindered a caravan.

The caravan could be kept close together only at the expense of marching speed. The German and African officers and NCOs were distributed along the column every 20 to 30 men, in such a way that those who lagged behind could more easily be made to catch up with their leader.

For security, each column was headed by an African NCO with four to ten men, who scouted as far as possible the way forward and to each side. One of the men at the head carried a machete, one an axe, and another some green twigs to place on the side paths and mark the trail, preventing stragglers from losing their way.

## Training of the *Askari*

### Individual Drill

Order arms, at ease and attention (the latter was described as being very difficult for the *Askari*, and was seen as one of the best discipline agents), natural march (as opposed to parade march), double march, turns (not easily done if the man was barefoot).

### Platoon Drill

Marching as one (the African march formation), forming two rows, parades, front march, halt, left and right turns and forming into rows.

### Company Drill

Column formation, movements, left and right swings, marching in line, attacking, running stride, halt, kneeling, company formation, company squad formation, movement in the same, shooters jumping out of ambush.

## Resolved Order

Swarming on all sides, movements, attack, melee and ambush.

## Rifle Drill

Aiming in all positions and target shooting at all distances: 50 – 100 metres, 200 metres, 300 – 400 metres. Sighting knowledge up to 400 metres (it was deemed impossible to teach the *Askari* distances above this, as they were thought incapable of learning higher numbers accurately enough). Firing single shots on command; firing salvos as a platoon. Loading, methods of holding the rifle (shoulder arms, present arms, order arms, sling arms and unsling arms), fix bayonet and bayonet thrusts and attacks.

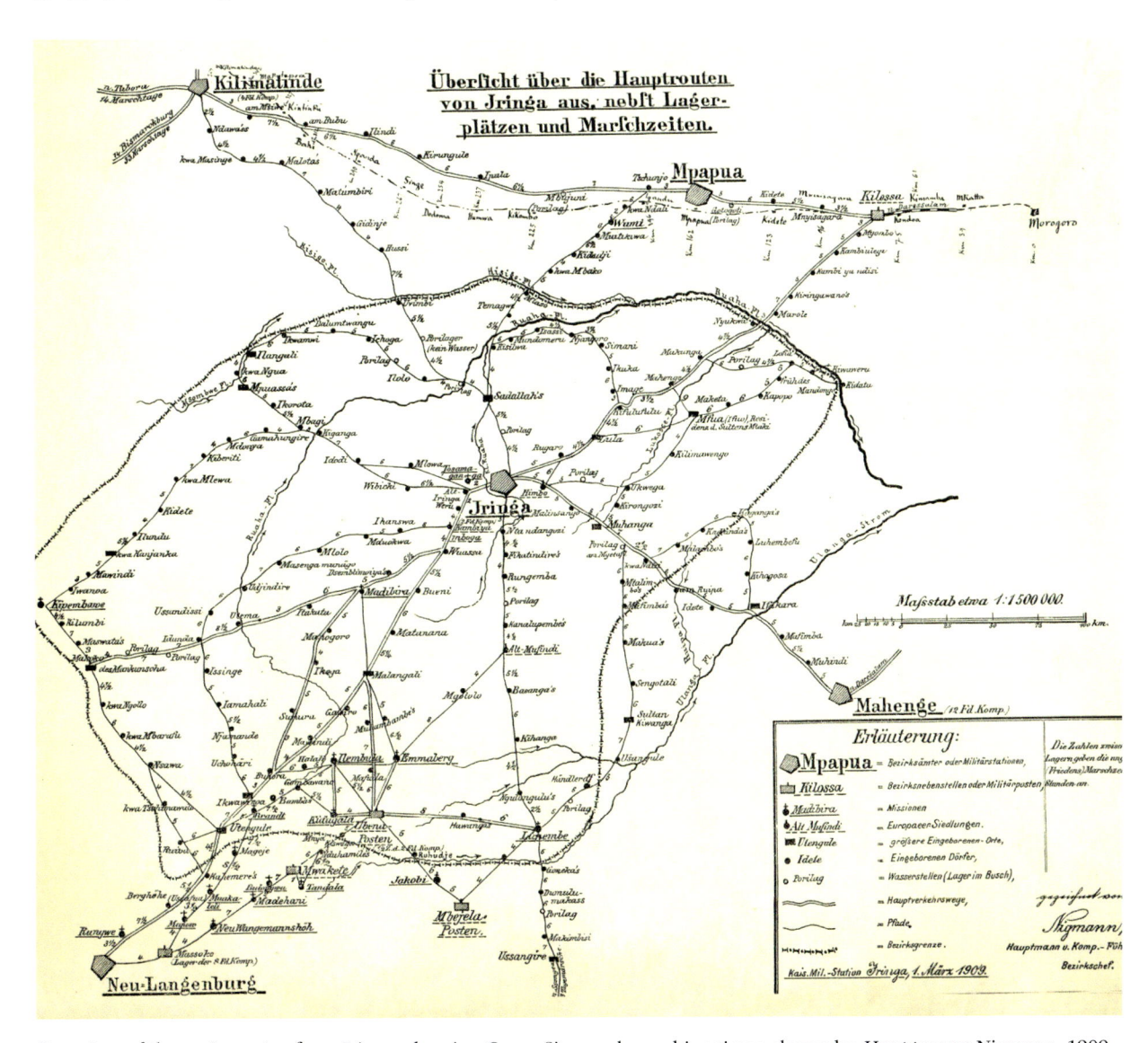

Overview of the main routes from Iringa, showing Camp Sites and marching times, drawn by *Hauptmann* Nigmann, 1909.

*Askari* after nine months in the field, 1915.

*Askari* Column fording a river.

*Askari* Company ready to march.

*Askari* in Rifle Training.

*Askari* Slow-Step Training.

*Askari* cleaning Rifles.

*Askari* on Exercise.

*Askari* on Parade with full Equipment.

*Askari* training with Machine Guns.

## Organisation of the *Schutztruppe* in 1914

On 8 February 1914, the East Africa *Schutztruppe* celebrated its 25th anniversary. At that time, it consisted of 68 officers, 42 medical officers, and about 200 military officials, NCOs, and medical NCOs. There were also about 2,500 *Askari,* including two officers and 180 NCOs.

The *Schutztruppe* was divided into fourteen **Field Companies** (*Feldkompanien*), each with about 160 *Askari*. In addition, an equally strong **Recruitment Depot** (*Rekrutendepot*), a **Signal Unit** (*Signal Abteilung*) of about 60 men (including eight-to-ten-year-old boys learning heliography) and a military band of 20 men were stationed in Dar es Salaam. Each company usually had around 40 to 60 porters, depending on local conditions. Once war broke out, this number rose to an average of 250 men or more, to whom the transport of food, ammunition and machine guns was entrusted.

The German personnel in each company usually consisted of a **Captain**, a **First Lieutenant**, one or two **Second Lieutenants**, a **Medical Officer**, a **Junior Paymaster**, a **Senior Sergeant**, a **Medical NCO** and two **NCO**s. Attached to the *Schutztruppe* commander's overall staff were five **Armourers**, three **Ordnance Personnel** and a **Saddler**.

There was also the 400-strong **Arab Corps** (*Araberkorps*), made up of men who volunteered for military service in wartime but which proved to have little combat value.

Each company had two or three machine guns. The following artillery were also distributed to the field companies: six old *C73* **Field Guns**, two **6 cm Colonial Guns**, and some **3.7 cm Quick-Firing Guns** and **3.7 cm Revolver Cannons**. Later, two **6 cm Naval Landing Guns** were brought on the *SS Rubens* and four **10.5 cm Field Howitzers** were brought out by the *SS Marie*. The ten **10.5 cm guns** of the cruiser *SMS Königsberg* were also added to the *Schutztruppe's* strength.

Returning from an Exercise.

The fourteen field companies were deployed as follows:

| 1st Company | Aruscha |
| --- | --- |
| 2nd Company | Iringa with an outpost at Ubena |
| 3rd Company | Lindi |
| 4th Company | Kilimatinde with an outpost at Singidda |
| 5th Company | Massoko with an outpost at Neu-Lagenburg |
| 6th Company | Udjidji with an outpost at Kassulu |
| 7th Company | Bukoba with an outpost at Ussui und Kifumbiro |
| 8th Company | Tabora |
| 9th Company | Usumbura |
| 10th Company | Daressalam |
| 11th Company | Kissenji with an outpost at Mruhengeri |
| 12th Company | Mahenge |
| 13th Company | Kondoa – Irangi |
| 14th Company | Muansa with an outpost at Ikoma |

## Uniforms, Equipment & Weapons of the *Polizeitruppe*

The *Polizeitruppe* did not belong to the *Schutztruppe*, but came under the authority of the colonial administration and provided security in districts where the *Schutztruppe* did not have a presence. Their tasks corresponded approximately to those of a gendarmerie. The police *Askari* were recruited from former *Schutztruppe* soldiers and thus had undergone proper military training. As a rule, older *Askari* of the *Schutztruppe*, who had lost some of their vigor during their years of service on strenuous expeditions but who were still capable of calmer service, were assigned to the *Polizeitruppe*.

The **Polizeitruppe Askari** wore the same uniform and equipment as the *Schutztruppe Askari*. It differed, however, by the addition of an oval white badge bearing a red letter 'P' on the left sleeve, the buttons on the tunic being made of brass, and the imperial eagle on the *Tarbusch* being larger than that of the *Schutztruppe*. Their armament was the same as the *Schutztruppe*.

## Organisation of the *Polizeitruppe* in 1914

The strength of the *Polizeitruppe* at the outbreak of war was as follows: four German Officers, 61 German Police NCOs, 147 African NCOs, and about 1,850 *Askari*. These were distributed in units of 60 to 120 men under the leadership of German Police NCOs in the various districts. The overall command was in the hands of a Captain with two Inspectors.

After the First World War began, the *Polizeitruppe* were transferred to *Schutztruppe* command. Following is a letter from *Schutztruppe Major* Langenn-Steinkeller to the German Resident Official at Gitega and the Resident Office Substation Usumbura, requesting a full inventory of their weapons along with their pay details, to facilitate the incorporation of the *Polizeitruppe Askari* into the *Schutztruppe*:

*"Schutztruppe"* Urundi Division (*Abteilung*),

Usumbura, 2 March 1915

178/Mob.

**A.) Resident Official, Gitega**

1) By order of command 5705/14 of 25 August 1914, the Urundi police department has been transferred to the *Schutztruppe*. The Urundi *Schutztruppe* must keep all *Askari* on their lists and is therefore asking to receive a list containing all *Askari* stationed in Gitega, Niakassu or any other highland area with their name, dog tags, depot and pay grade. These persons are listed in the local list and in the earnings and subsistence reports as assigned to the residency Gitega (or Niakassu) and receive their pay at the expense of the M fund. In the pay books of the *Askari* to be transferred to the Urundi *Schutztruppe*, "25.8.14 assigned to the Resident Official Gitega" should be written. Accordingly, the pay lists in Gitega, etc., are to be updated and certified copies sent to the Urundi *Schutztruppe*.

2) With the transfer of the police units to the *Schutztruppe*, the entire inventory of clothing, equipment, weapons and ammunition has also been placed under the administration of the Urundi force. Therefore, I ask you to supply a list of:

(A) the remaining stock of clothing & equipment

and

(B) the existing stocks of rifles (with *M98*, *M71* with ejector [*M71/84*], *M71* without ejector, and carbines *M98* and *M71* separately), bayonets (with steel scabbards, leather scabbards and without scabbards), revolvers, pistols and ammunition (separating *M71* revolver and pistol ammunition). Those in the hands of Europeans and *Askari* should all be included and marked as such.

The items sent by the *Schutztruppe Abteilung* Urundi to the headquarters at Gitega for Gitega's defence should not be counted.

From here I will decide which items to keep current and in alignment and which weapons and ammunition are available there and may be used at their sole discretion. However, I ask you to take the utmost care to monitor all usage and to handle it as sparingly as possible.

3) Policemen hired at Gitega do not belong to the Urundi *Schutztruppe*; their pay therefore comes from the S fund. It is in the interest of the *Schutztruppe* to train the *Polizeitruppe* up to their own standard. I will readily approve the appointment of an *Askari* to the *Schutztruppe* as soon as the required level of training, primarily in shooting, is achieved. Upon approval, a message will be sent on the day of hiring, with the *Askari*'s name and depot, following which their pay book and dog tag will be sent.

4) I hereby request a dutiful declaration that, apart from those in the requested lists, there are no further stocks.

**A.) Resident Official Substation, Usumbura.**

1) For your information.

2) Seven *Askari* from the *Schutztruppe Abteilung* Urundi shall be seconded for the administration of the Resident's district. As with the personnel at Gitega, Niakassus, etc., these are listed in the main list of the Urundi Division but are paid from M. funds. I request you to set up the pay lists there and also to submit transcripts.

New uniforms, when needed, shall be requested from the *Schutztruppe* depot. Each man in Usumbura is to have 10 cartridges on hand. The bread bags of these men can carry another 110 rounds plus a bandage. Crates of 1,500 *M71* cartridges shall be carried separately in the field. This ammunition is considered a reserve for the defence of the *Boma* at Usumbura, for which the above-appointed men – and in addition the left-behind recruits, the sick, the Arabs, and the police – are to be used.

3) For the pay of police serving in the Urundi *Schutztruppe*, those named on the lists are added to the budget of the *Schutztruppe* department as *Ruga-ruga* and are paid out of M. funds. The rest of the police are therefore at the disposal of the Resident's sub-station, so should receive their wages from S. funds. As to their military training, I am most indebted to my colleague in Gitega for his orders.

4) The seven Baharia sailors of the whale boat are within the budget of the *Schutztruppe*, but are considered as assigned to the Resident's sub-station.

signed Langenn (*Major*)"

*Editors Note: M. funds and S. funds must be assumed from the context to mean respectively Schutztruppe and Colonial Government funds.*

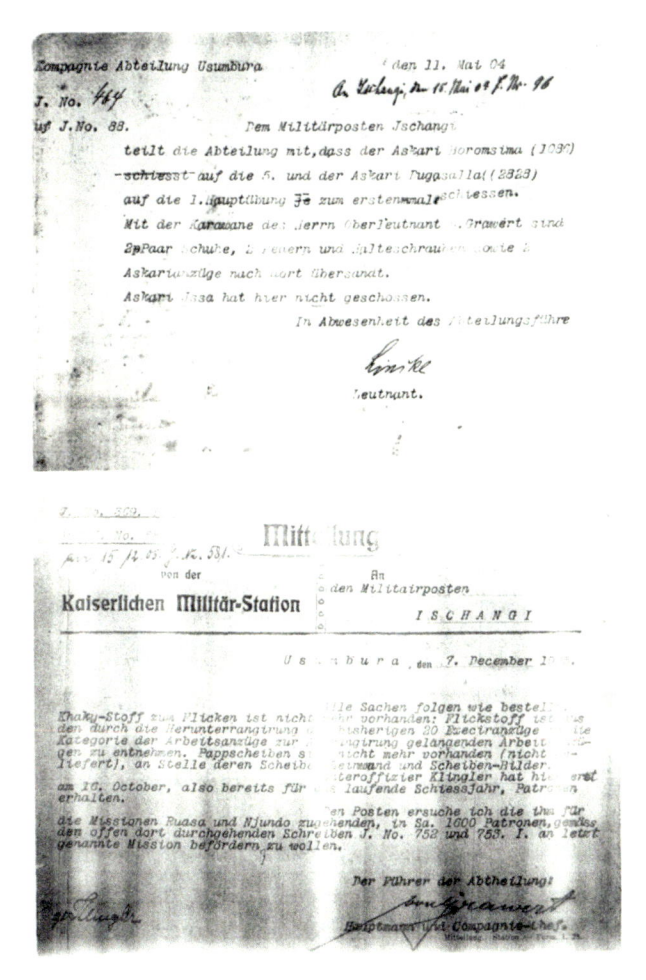

Messages to the Military Post at Ischangi.

Lieutenant Colonel Lettow-Vorbeck in 1913. Alhough appointed Commander of the Kamerun *Schutztruppe*, he was transferred to German East Africa instead.

Standing at Attention.

The Bugler gives the Signal.

*Polizeitruppe Askari* on Guard Duty.

Scaling an Obstacle Course.

| Standort | Truppenteil | Deutsche | Askari | Ruga-ruga | Masch.-Gewehre | Geschütze |
|---|---|---|---|---|---|---|
| Tanga | 15., 16., 17. F.-Ko., 4. S.-Ko.*) | 145 | 700 | 50 | 11 | — |
| Tanga | Küstenschutz | 44 | — | — | — | — |
| Uschiwa, Rombo | 3., 9., 11., 13. F.-Ko. | 184 | 825 | — | 10 | 2 |
| Taveta | | 86 | 561 | 46 | 5 | — |
| Neu-Moschi | 4. F.-Ko., 5., 6. S.-Ko. | 170 | 416 | — | 7 | 2 |
| Kware | 8. F.-Ko. | 16 | 176 | — | 2 | — |
| Engare Nairobi | 10., 21. F.-Ko., 8., 9. S.-Ko. | 149 | 364 | — | 4 | — |
| Aruscha, Old. Sambu, Aruscha, Engereta | | 50 | 109 | 82 | — | — |
| Bezirk Muanza | 14., 26. F.-Ko., ½ 1. S.-Ko. | 143 | 380 | 600 | 3 | 6 |
| Bezirk Bukoba | 7. F.-Ko., ½ 1. S.-Ko. | 61 | 273 | 520 | 4 | 3 |
| Ruanda | Abteilung Tanganjika, 22. F.-Ko., Komp. Möwe, Urundi-, Bismarckburg-Abteilung | 20 | 105 | — | 2 | — |
| | | 219 | 580 | — | 4 | — |
| Langenburg | 5. F.-Ko., ½ 2. S.-Ko. | 38 | 386 | — | 6 | 1 |
| Mahenge | 12. F.-Ko. | 13 | 363 | — | 2 | — |
| Iringa | ½ 2. F.-Ko. | 6 | 209 | — | 1 | — |
| Lindi | 20. F.-Ko., Rugaruga-Komp., Polizei | 17 | 219 | 174 | 1 | — |
| Küstenschutz Kilwa—Lindi | | 27 | 62 | 48 | — | — |
| Rufidji | Delta-Abteilung | 150 | 42 | 58 | 4 | 2 |
| Daressalam | 18., 23., 24., 25. F.-Ko., 10. S.-Ko., Landsturm | 279 | 537 | — | 9 | 3 |
| Bahnschutz Mittellandbahn | | 80 | — | — | — | — |
| Morogoro | Wachkommando, Rekr.-Depot Tabora | 7 | 301 | — | — | — |
| Küstenschutz Bagamojo | | 14 | 44 | 231 | — | — |
| Küstenschutz Pangani | | 5 | 5 | 94 | — | 2 |
| Bahnschutz Nordbahn | | 47 | — | 41 | — | — |
| Wilhelmstal | Landsturm, Rekr.-Depot | 31 | 112 | — | — | — |
| Muhesa | Rekr.-Depot | 5 | 302 | — | — | — |
| Moschi | Grenzschutz | 12 | 20 | — | — | — |
| | | 2008 | 7091 | 1894 | 75 | 21 |

*) F.-Ko. = Feldkompagnie, S.-Ko. = Schützenkompagnie.

A listing of the Organisation and Strength of the *Schutztruppe* as it stood in March 1915. The first column lists the Garrison Towns, the second notes the Units based there, and the other columns tally the number of Germans, *Askari, Ruga-ruga* (Irregulars), Machine Guns and Artillery pieces assigned to each Garrison.

Outside the Barracks in Dar-es-Salaam.

Forward March.

*Askari* in Action, painted by W. Kuhnert.

*Askari* Soldier leading a Column of Bearers, painted by W. Kuhnert.

# Imperial German Colonial Troops & Police in Africa

**I**

**II**

Two pages from *Hauptmann* Nigmann's
German East Africa photo albums. Nigmann
can be seen in Photo I as number 4,
and in Photo III as number 1.

Note that several *Askari* wear the Gun Layer's
Badge on the upper left arm, and at least one wears
Markmanship Bands on his cuffs.

**III**

## I

1. Kav.-Uuffz. Hellwig
2. Feldwebel Hermann
3. Leutnant Arneth
4. Hauptmann Nigmann
5. Unt.-Zahlm. Stops
6. Sergeant Birkner
7. Schautsch Ali Nur
8. Ombascha Kilima
9. Lottschautsch Mabruk Hussein
10. Ombascha Ngamia
11. " Bagamoyo
12. " Hassan
13. " Chitima
14. Gefr. Mohamed Achmed
15. Ombascha Nguna
16. " Selimani
17. " Osmani
18. " Kofia mbaya
19. Schautsch Asmani
20. Ombascha Majuta
21. Lottschautsch Munipembe
22. Ombascha Shabaha
23. " Ali Usango
24. " Saidi
25. Schautsch Bandagoha
26. Ombascha Gombagena

## II

1. Schautsch Bandagoha
2. Ombascha Majuta

## III

1. Hauptmann Nigmann
2. Oberleutnant Schimmer
3. Unteroffizier Reupke
4. Gefr. Mohamed Achmed
5. Ombascha Ngamia
6. Lottschautsch Mabruk Hussein
7. Ombascha Tanganyika
8. Unteroffizier Birkner
9. Lottschautsch Munipembe
10. Ombascha Shabaha
11. " Nguna
12. " Mohamed Ali
13. " Bagamoyo
14. " Saidi
15. Schautsch Chitima
16. Ombascha Idi
17. Schautsch Ali Nur
18. Ombascha Ali
19. " Ngombagena

Map of *Schutztruppe* Garrisons in East Africa.

## The First World War in German East Africa

On 5 August 1914, Governor Schnee declared a state of war in the colony and issued a call to arms; so began the longest campaign of the entire war.

By March 1915, after the call-up of reservists and the incorporation of the *Polizeitruppe*, there were 2,008 Germans and 7,091 *Askari* in *Schutztruppe* service. This number included about 800 recruits and 1,894 irregulars or *Ruga-ruga*, who could best be described as armed bearers, though over the course of the war they were sometimes absorbed by regular field companies. At the peak of their strength in late 1915, the *Schutztruppe* mustered around 3,000 Germans and 11,000 *Askari*.

Initially, the *Schutztruppe* held their borders intact and defeated a British Indian army in its attempted invasion at Tanga. In 1916, the allies launched simultaneous offensives with a South African and British force invading from the north-east, British Royal Naval landings along the coast and the Belgians advancing from the Congo. Gradually the Germans were beaten back.

In October 1918, the *Schutztruppe* marched through the districts of Iringa and Lagenburg and crossed the German-Rhodesian border at Fife. On 12 November 1918, they fought their last battle of the campaign, north of Kasama. The following day the German force received the first accurate news of events in their homeland. On 14 November while at Kasama, their commander von Lettow-Vorbeck received orders from Berlin for a ceasefire, after a four-and-a-half-year campaign. At this time the *Schutztruppe* had 30 officers, 125 non-commissioned officers, 1,168 *Askari*, 1,522 porters and a few hundred female camp followers. German East Africa was then placed under the administration of a British Mandate.

Upon his return to Germany, *General* von Lettow Vorbeck inspects an Honor Guard at the Lehrter Train Station, March 1919.

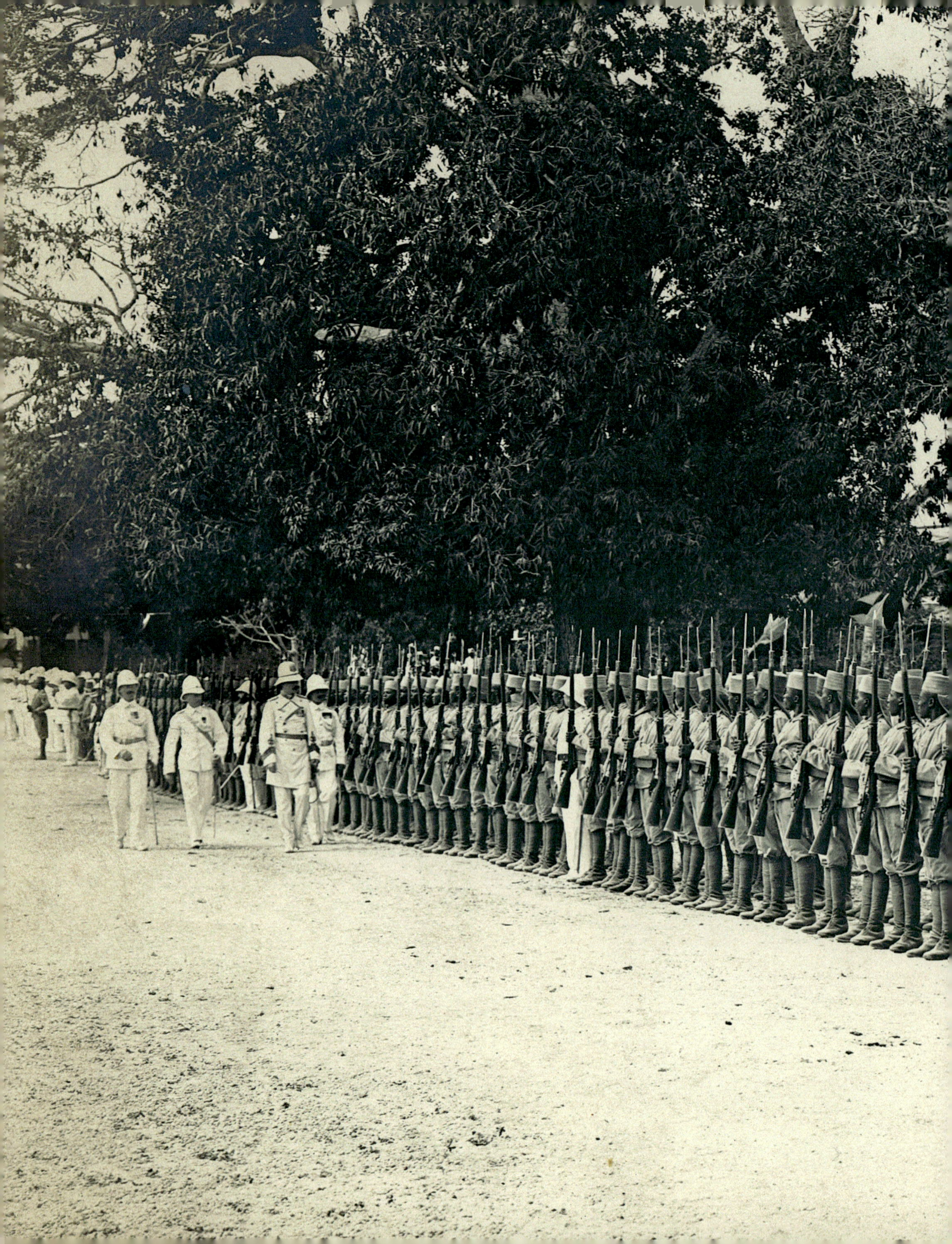

*Askari* inspected by the Governor of German East Africa.

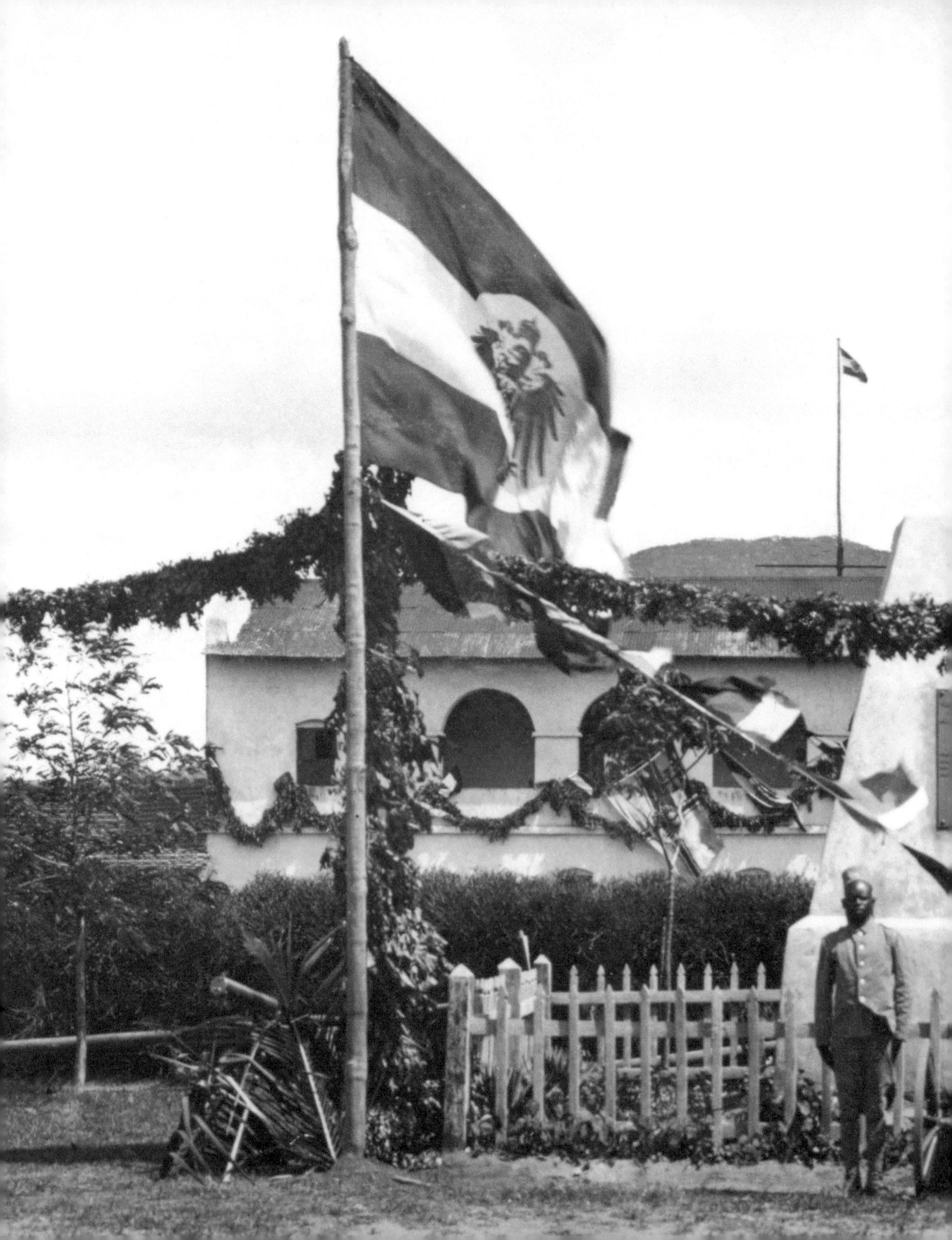

*Askari* at the Iringa Memorial to the fallen in the Maji-Maji Rebellion. Note the 3.7 cm Quick-Firing Gun and two Machine Guns.

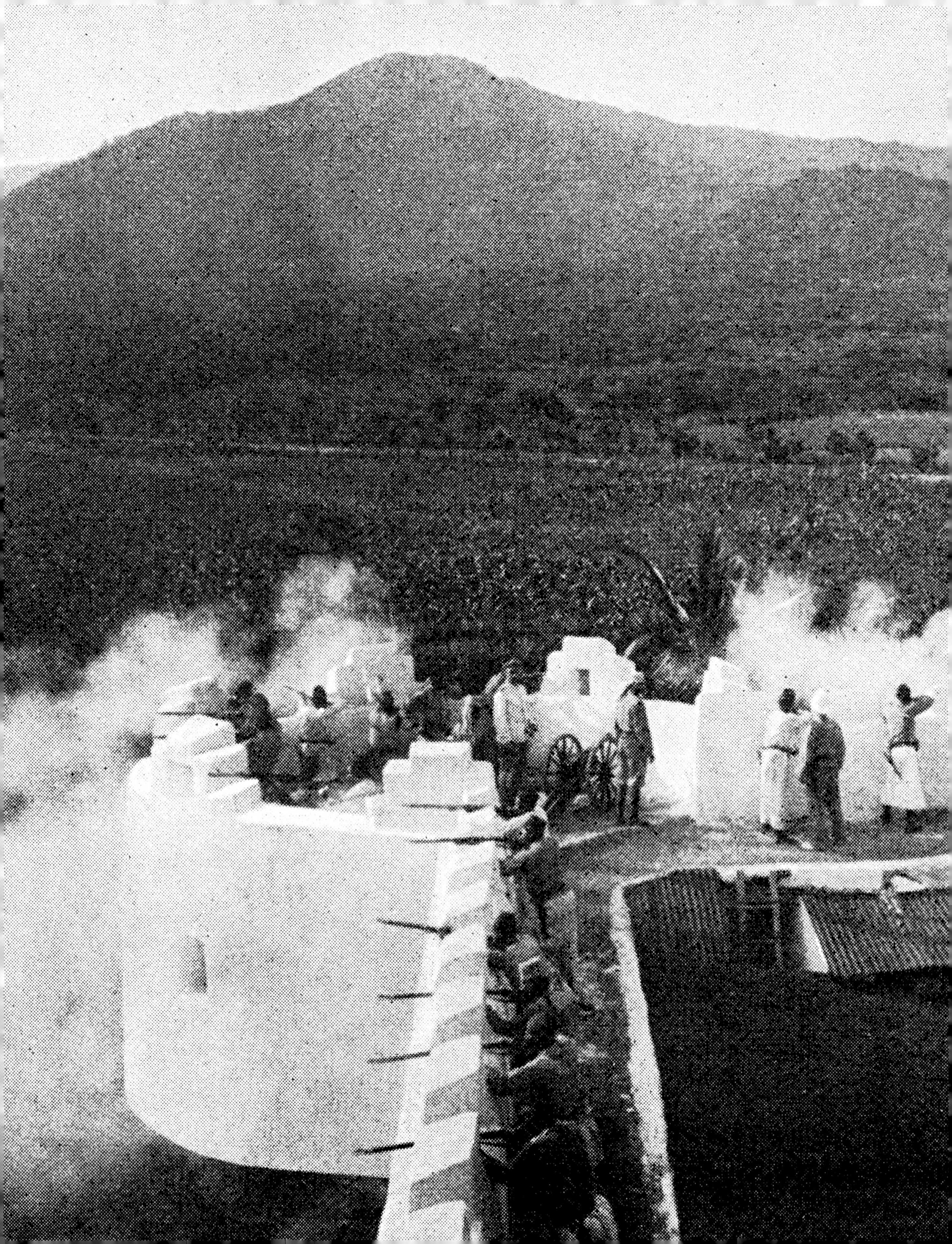

*Askari* practice defending a Fort.

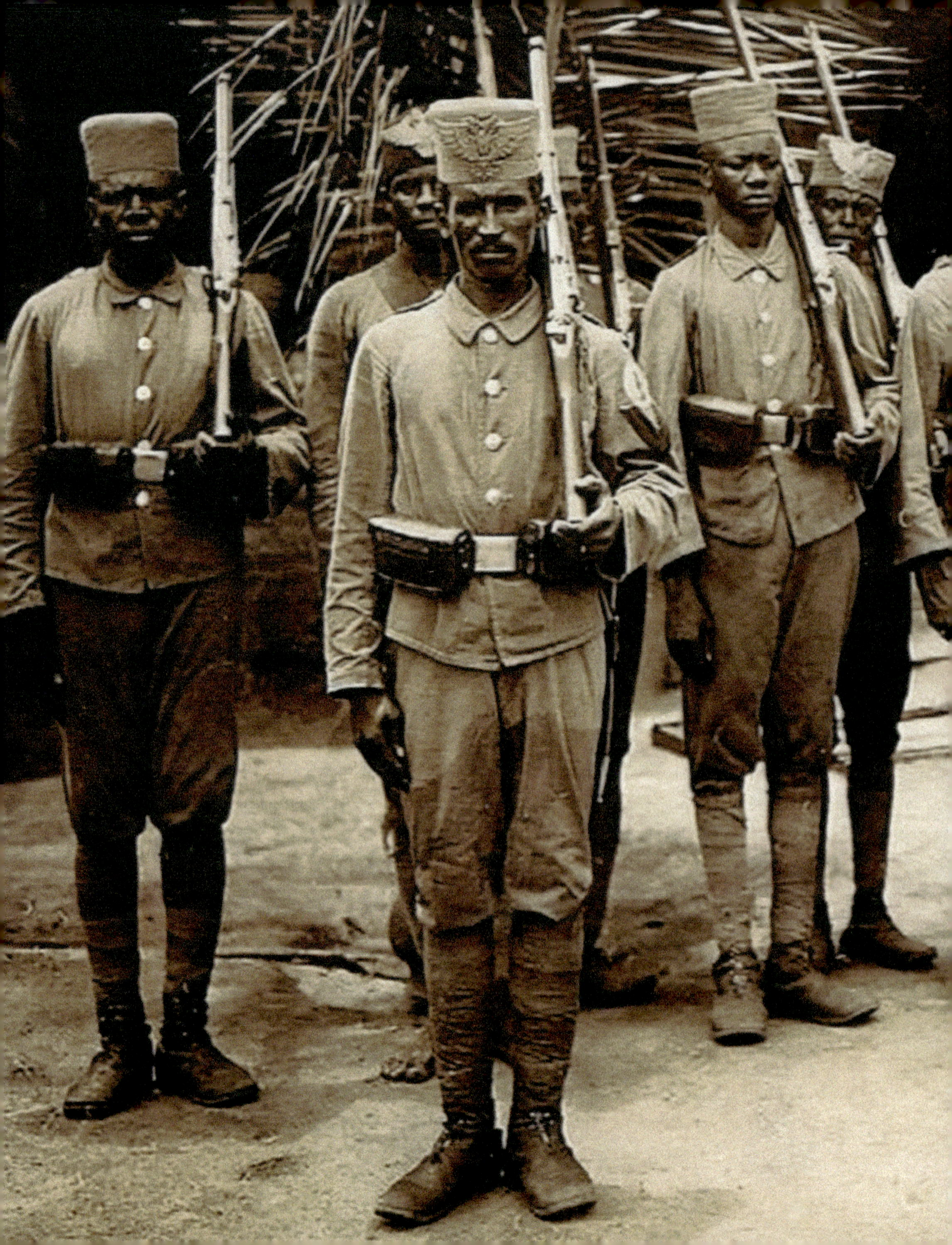

*Askari* Roll Call.

*Schutzruppe* Officers and *Askari* attend a briefing at the beginning of the First World War.

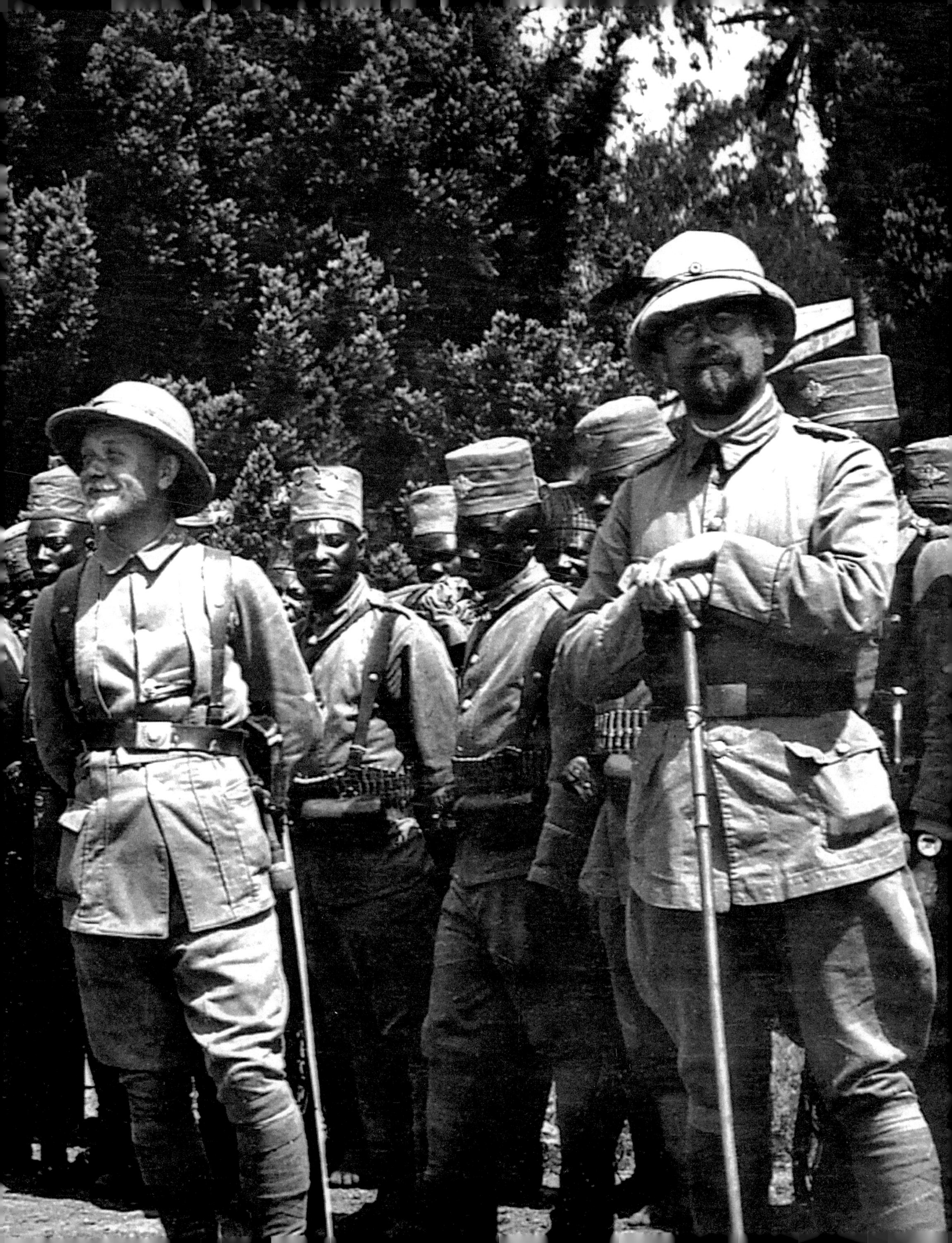

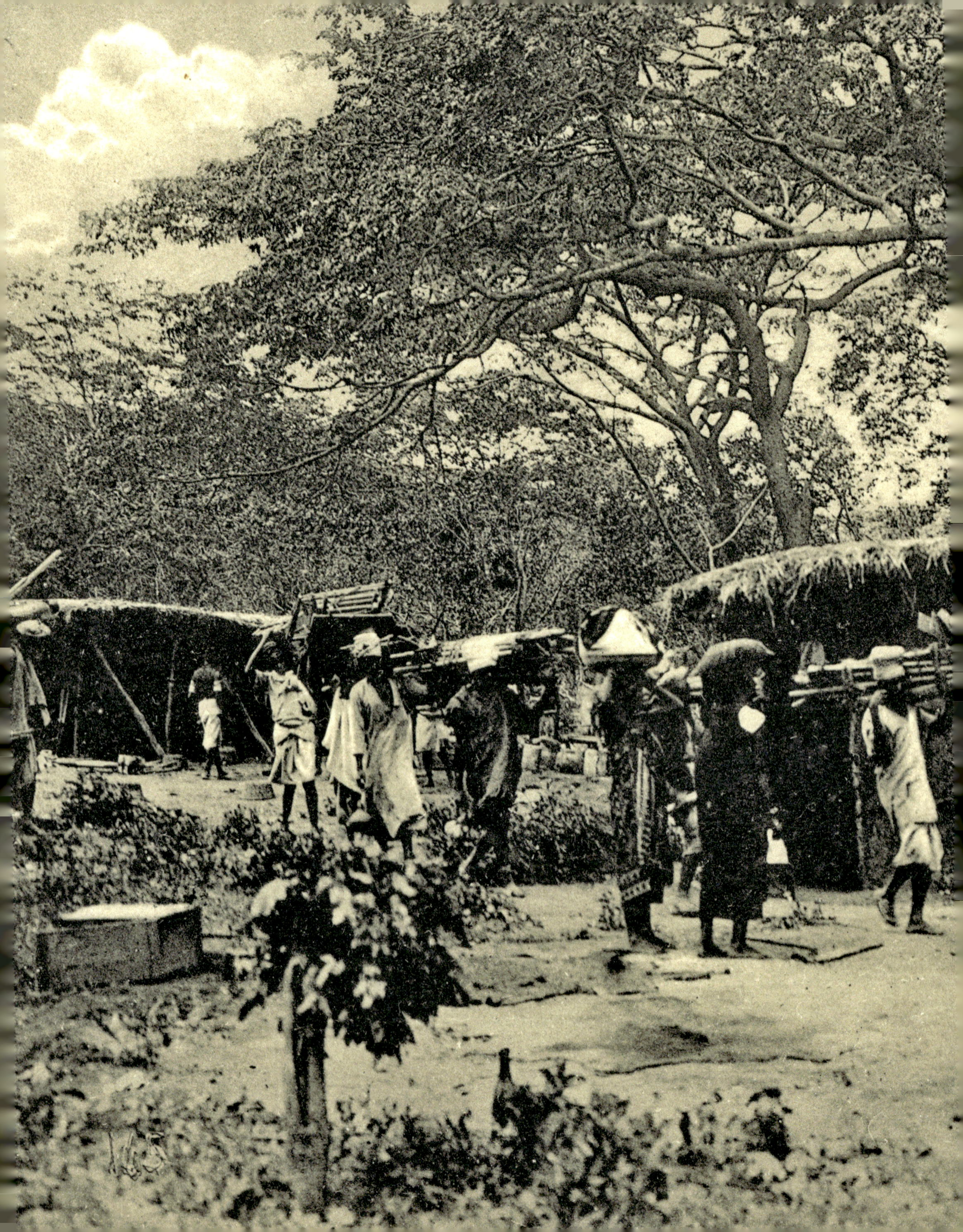

D.-O.-Afrika
Eine Trägerkarawane.

*Schutztruppe* Column fording a river in German East Africa.

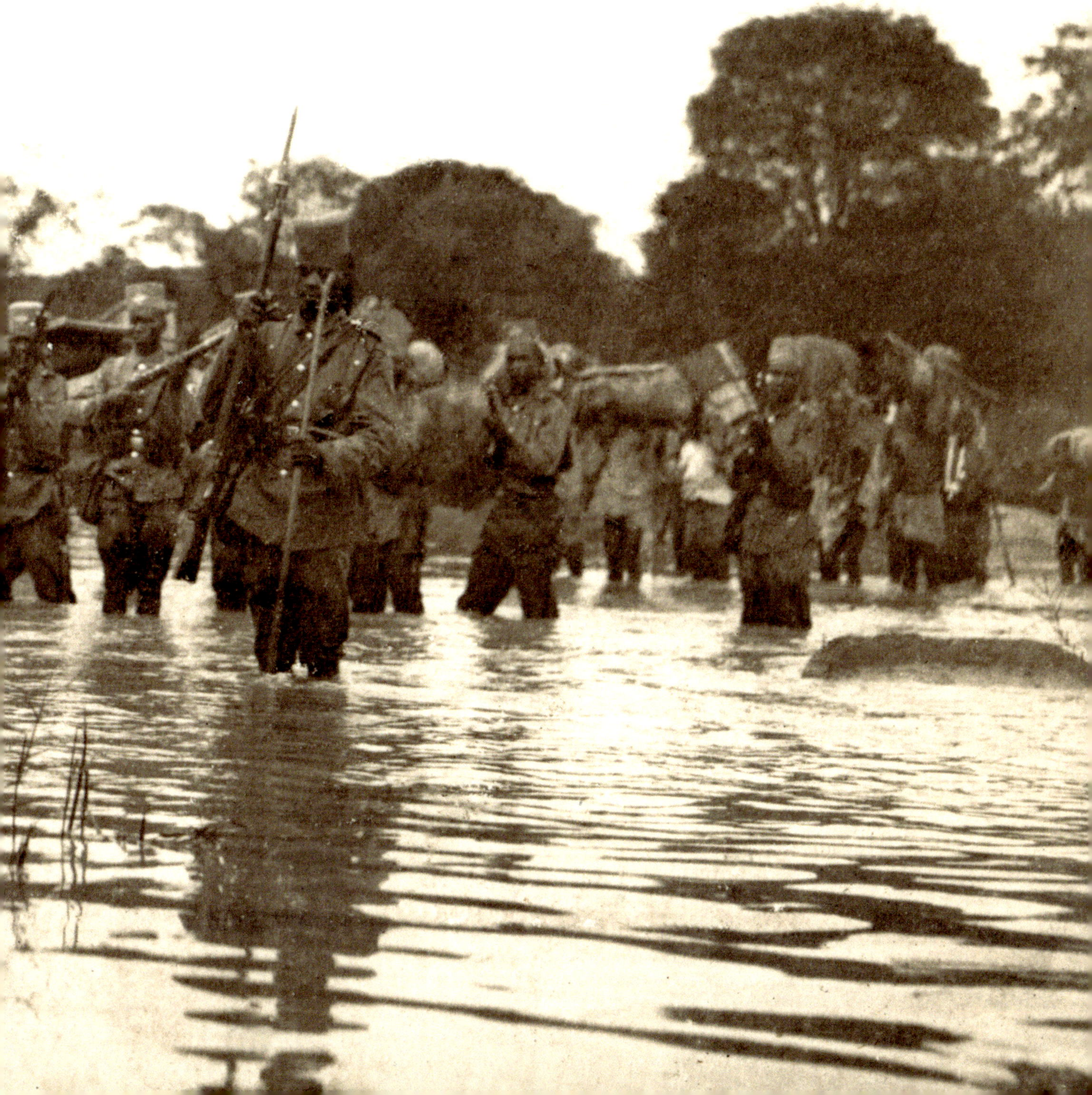

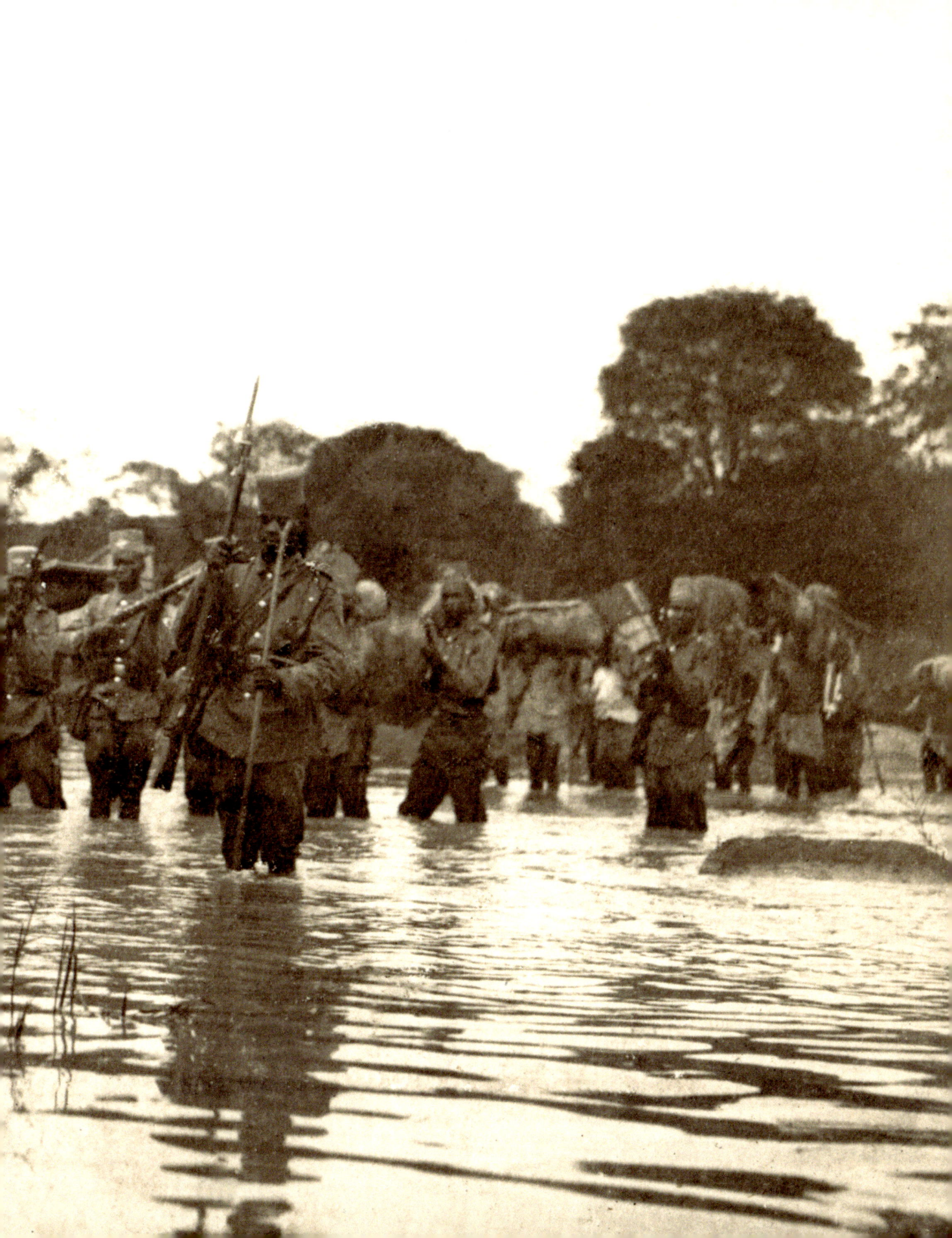

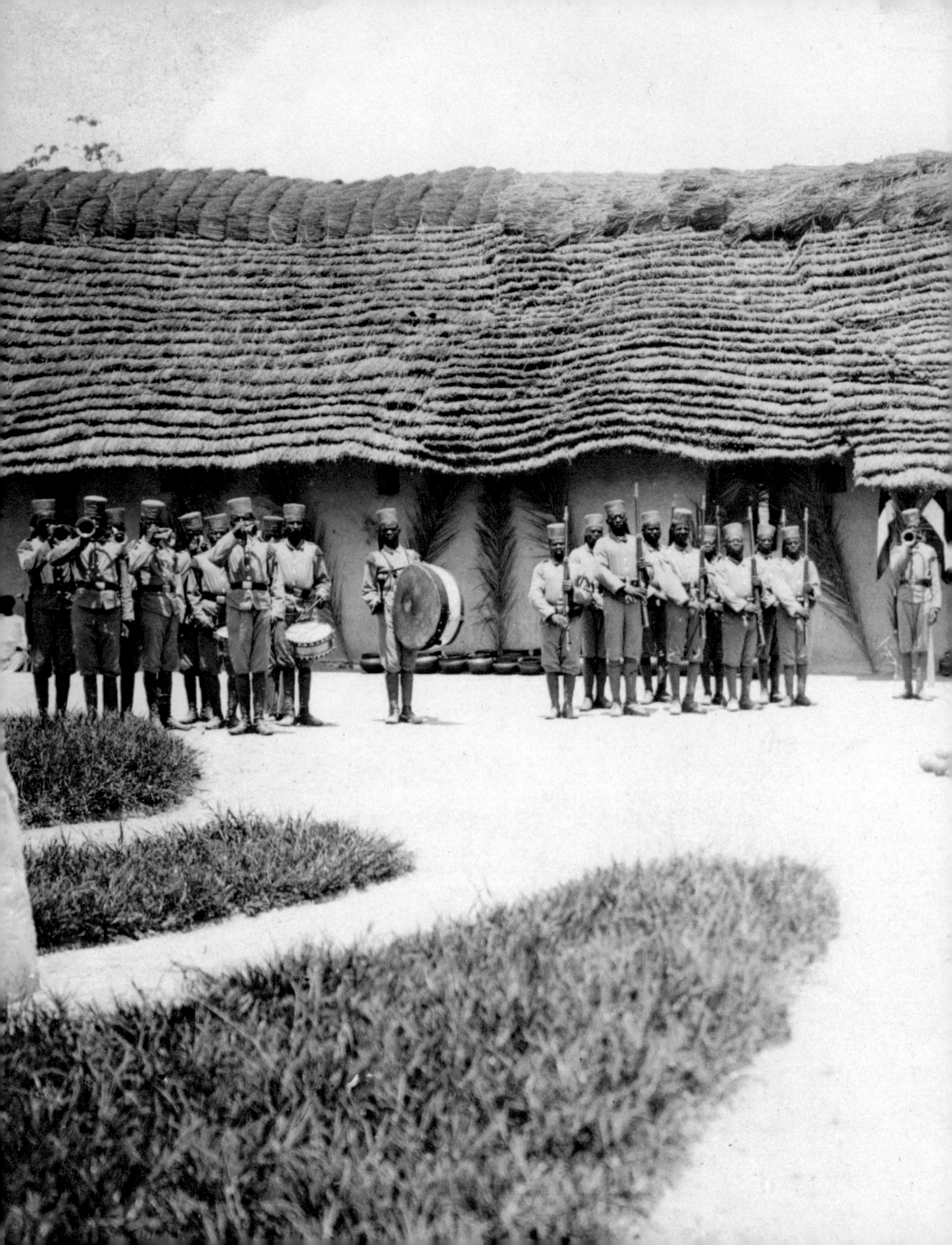

*Askari* Band and Troops at Tabora (and diorama of same by the author).

*Schutztruppe* Band in Dar-es-Salaam, playing Fanfare. On inset Postcard, the other side of Trumpet Banners can be seen.

Deutsch-ost-afrikanischen
Glückwunsch

sendet zum

African Civilians and Usambara Railway workers, with two *Askari* Policemen in the center.

Photo: Gebrüder Haeckel, Berlin.

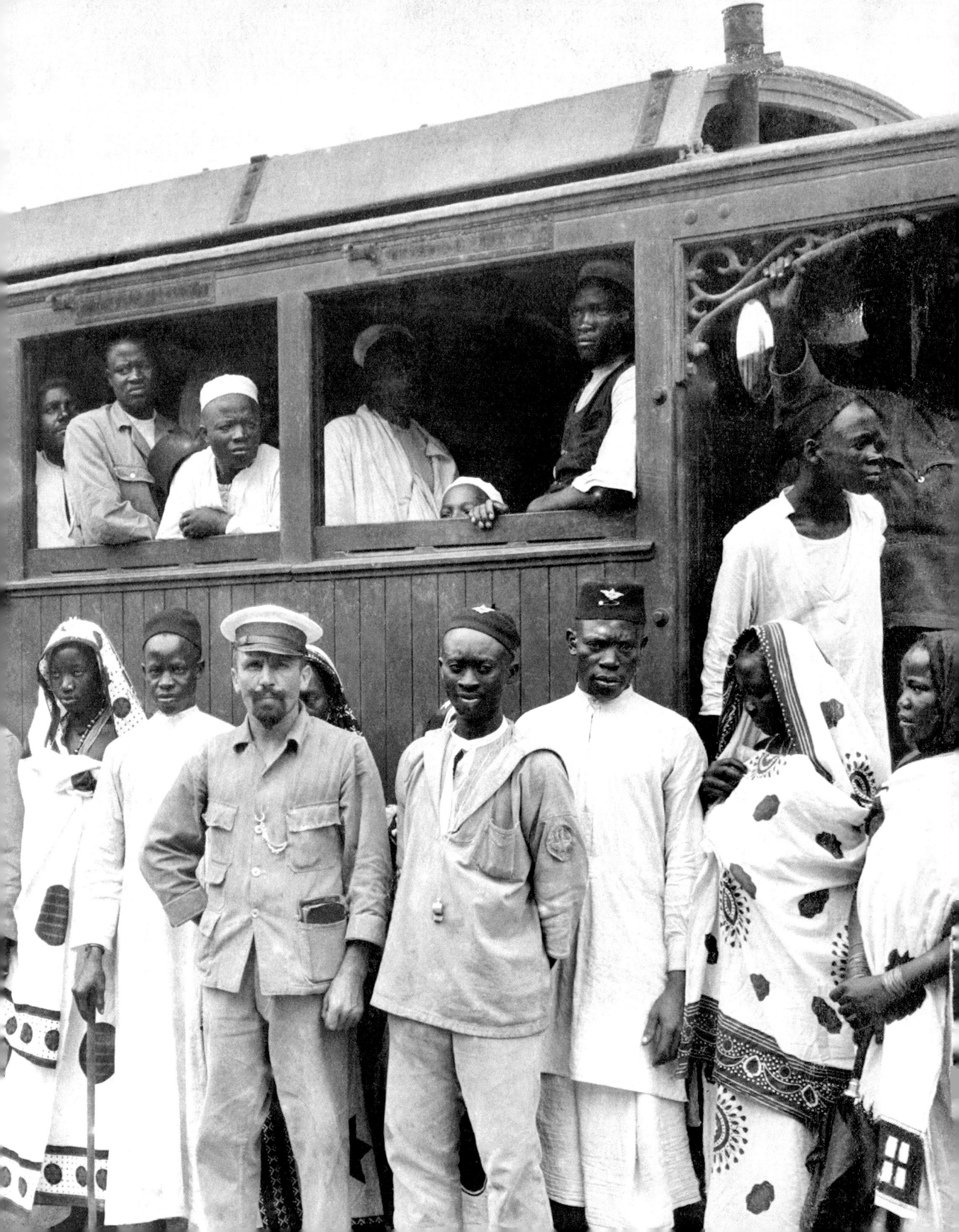

Photo: *Süddeutsche Zeitung* / Alamy Stock Photo

German East African *Polizeitruppe* at a Station on the Usambara Railway.

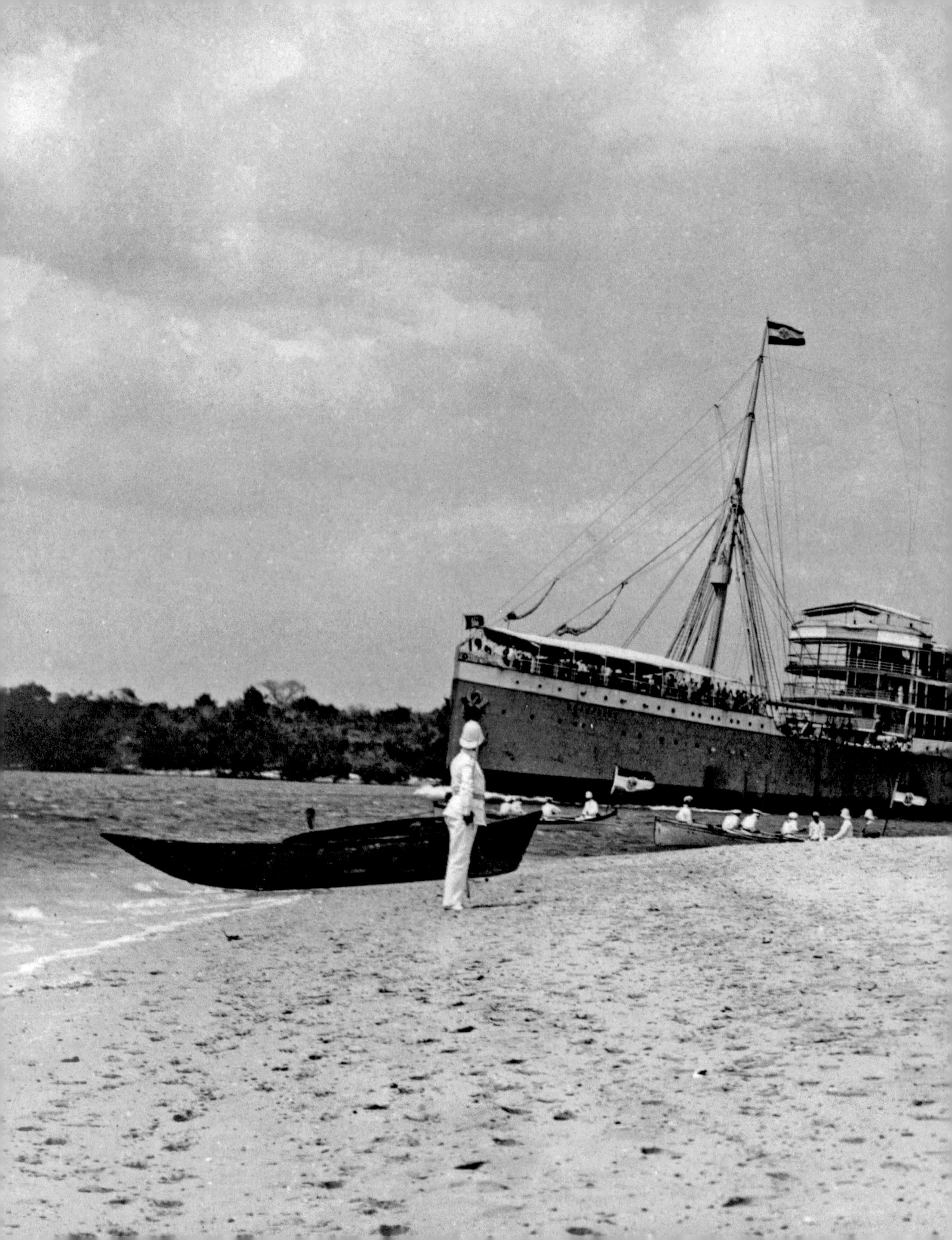

*Schutztruppe* stand at attention as the Governor's Steamship departs Dar-es-Salaam.

Local newspaper with reports from the African and European theaters, 31 December 1915.

## Heia, Safari

*Schutztruppe* Anthem

Lyrics by Anton Aschenborn,
set to music by Robert Götz

| | |
|---|---|
| How many times have we walked | Wie oft sind wir geschritten |
| On the narrow native path? | Auf schmalem Negerpfad |
| Surely, through the middle of the steppes, | Wohl durch der Steppen Mitten, |
| When the morning approaches early, | Wenn früh der Morgen naht; |
| How we listen to the sound | Wie lauschten wir dem Klange, |
| Of the old trusty songs | Dem alten trauten Sange |
| Of the carrier and *Askari*: | Der Träger und Askari: |
| Heia, heia, safari! | Heia, heia, Safari |
| | |
| Over steep mountain and clefts | Steil über Berg und Klüfte |
| Through deep jungle at night, | Durch tiefe Urwaldnacht, |
| Where muggy and humid is the air | Wo schwül und feucht die Lüfte |
| And the Sun never laughs; | Und nie die Sonne lacht, |
| Through waves of steppe grass | Durch Steppengräserwogen |
| We moved across | Sind wir hindurchgezogen |
| With carriers and *Askari*: | Mit Trägern und Askari: |
| Heia, heia, safari! | Heia, heia, Safari! |
| | |
| And we sat by the fire | Und saßen wir am Feuer |
| At night before the tent, | Des Nachts wohl vor dem Zelt, |
| Lying as in quiet celebration, | Lag wie in stiller Feier |
| Around us the next world. | Um uns die nächt'ge Welt; |
| And from over the dark slopes | Und über dunkle Hänge |
| It sounds like a distant tone | Tönt es wie ferne Klänge |
| From carriers and *Askari*: | Von Trägern und Askari: |
| Heia, heia, safari! | Heia, heia, Safari! |
| | |
| As I start the final journey, | Tret' ich die letzte Reise, |
| The big voyage at last, | Die grosse Fahrt einst an, |
| Sing me up with that tune | Auf, singt mir diese Weise |
| Instead of sad songs then, | Statt Trauerlieder dann, |
| So that to my hunter's ears, | Dass meinem Jaegerohre |
| There at heaven's gate, | Dort vor dem Himmelstore |
| It sounds like a *Halali*:* | Es klingt wie ein Halali: |
| Heia, heia, safari! | Heia, heia, Safari! |

* *Editor's Note: Halali is a Swahili word roughly translated as a validation or permission.*

# Map

*Maschena · Birni · Kuka · Tsad See 250 · Ftu...*
*Komadugu*
*Katagum · B O R N U*
*Gudjeba · A · Logon · BAGIRM*
*D · Gongola · Masseny*
*O K O T O · Baldao · Ba Busso*
*Badiko · Jakubu · 1800 · Logon*
*Bautschi · Mindif B.*
*A U S S A · Muri · Jola · Garua*
*Keffi · Benue · S T A A T E N · Lame · Lai*
*Loko · A*
*Wukari · D*
*Takum · Gendero-B. · Ngaundere · Benue · F R A N*
*△ 2700 · A*
*Banyo · M · C O N*
*Cross R. · Ngambe · Mbam · Tibati · N*
*scha · Ndjerem · Jaunde · A*
*Town · KAMERUN · Gasa*
*Ngaundere · Joko*
*Albrechts · Lom*
*Mangamba · GEBIET*
*Kamerun · Jaunde Stat.*
*o Poo · Edea · Nyong · Benia*
*Victoria · Lolodorf · Kadei*
*Lobetha · Marie*
*Plantation*
*Kribi · Fan*
*fra · Gr. Batanga*
*Bay · Campo*
*Batta · Ssanga*
*Eyo · (Franz.) · Ngoko*

# Part III
# **Cameroon**
## *(Kamerun)*

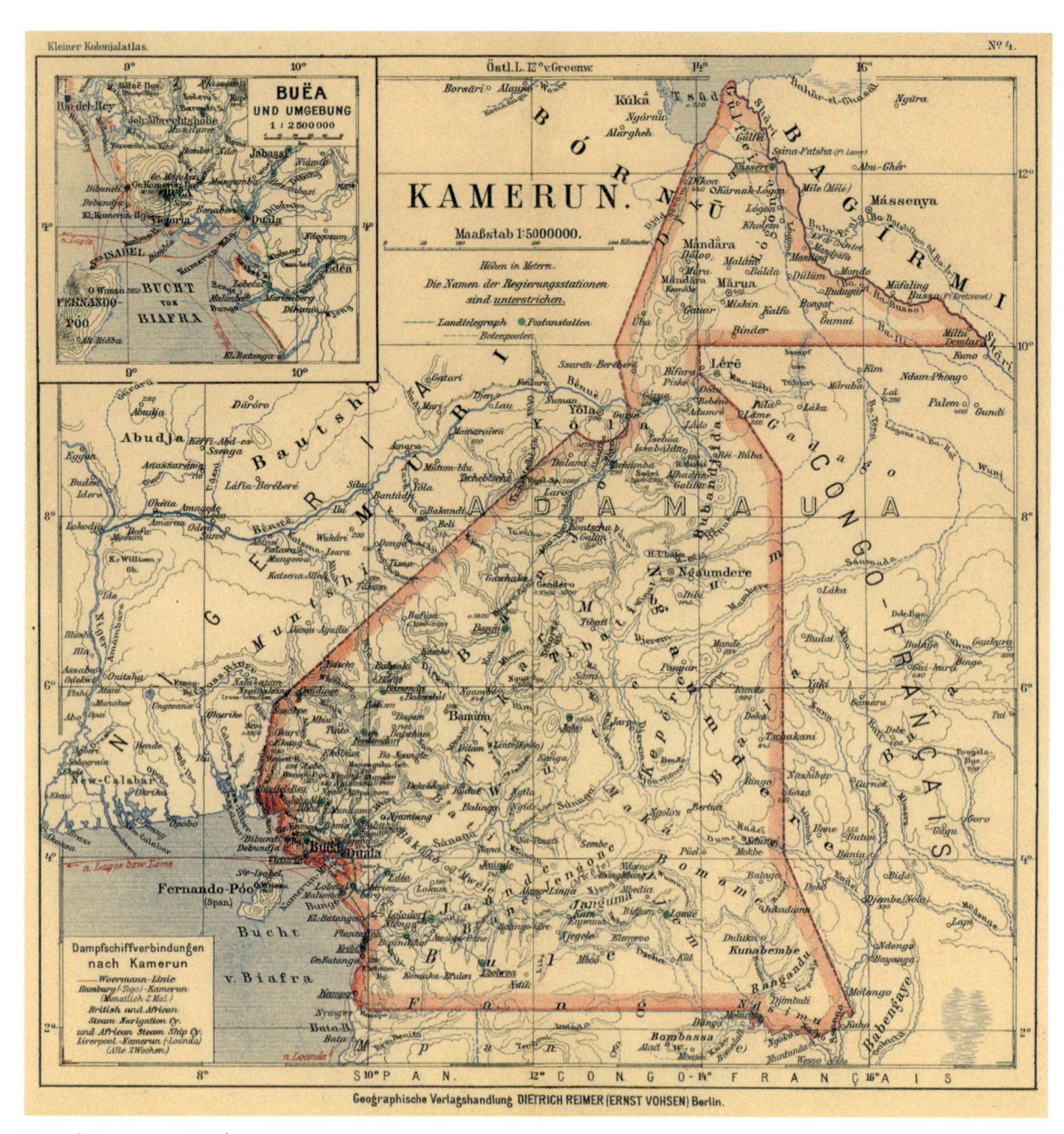

Map of Cameroon, 1906.

## Commanders of the *Schutztruppe*

1894 – 1896   *Hauptmann* Max von Stetten

1897 – 1901   *Major* Oltwig von Kamptz

1901 – 1903   *Oberst* Kurt Pavel

1903 – 1908   *Generalmajor* Wilhelm Mueller

1908 – 1913   *Oberstleutnant* Harry Puder

1914 – 1916   *Major* Carl Zimmermann

## Formation of the *Schutztruppe* and *Polizeitruppe*

Initially, the Governor had only a small number of African policemen available, a total of 54 men in 1889. These were intended to increase the government's reputation with the indigenous people and to undertake smaller expeditions to the interior. This was until the colonial department approved a request on 30 October 1891 for the formation of a *Polizeitruppe* of surplus men from the Gravenreuth Expedition. From this day, the real story of the Cameroon *Schutztruppe* began.

In 1893, the *Polizeitruppe* stood at two Europeans and 140 African soldiers, made up of men from Sierra Leone, Togo, Liberia, and Dahomey. The latter were purchased slaves, who had to earn back their purchase price before they received the same monthly salary of thirty Marks as the other recruits. This injustice and other grievances caused the Dahomey soldiers to mutiny on 15 December 1893. During the ensuing two-day battle, the mutineers even made use of the artillery guns in their possession. Fortunately for the Germans, the West African station ship S.M.S. *Hyäne* was able to quell the armed uprising.

L to R: Seals of the Governor of Cameroon and Cameroon *Schutztruppe* Commander.

The Dahomey Mutiny had clearly shown that the Organisation and strength of the *Polizeitruppe* force was insufficient. Therefore more Sierra Leone and Wey natives were recruited. In addition, at the behest of *Hauptmann* Morgen in Cairo, who had a deep insight into the conditions in Cameroon, the colonial administration hired a number of Sudanese fighters who had previously served in the Anglo-Egyptian forces. *Leutnant* Dominik and *Feldwebel* Krause, both of whom had also served in the Anglo-Egyptian army as well as the *Wissmanntruppe* in East Africa, arrived with the Sudanese reinforcements on 12 April 1894, accompanied by *Hauptmann* Morgen.

This constituted the only such recruitment of the Sudanese, as these "Sons of the Steppe" could not tolerate the humid climate of Cameroon. At this point further recruiting of the Wey took place.

The *AKO* which officially ordered the transformation of the *Polizeitruppe* into a *Schutztruppe* arrived a few weeks later. The expansion of the force, however was very slow. While

# Schutztruppe für Kamerun

**Paymaster**

wearing the
White Tropical Uniform

**Junior Paymaster**

wearing the
White Tropical Uniform

**Ordnance NCO**

wearing the
White Tropical Uniform

**Medical Officer**

wearing the
Grey Home Uniform

**und Togo.** 1896/1897

Shown here are uniforms approved for German Schutztruppe personnel in Cameroon and Togo on 19. November 1896, and for home service on 11. March 1897. Uniforms with red facing colors were authorized for both colonies in 1896, but in 1912 Togo adopted yellow uniform facing colors. The artist most likely used printed regulations to envision various changes in uniform, and may not have seen actual uniforms. Therefore some uniforms depicted may be theoretical. For example, there were no German Lance Corporals in Cameroon or Togo, and Togo had very few Schutztruppe personnel at all. The artwork also mistakenly shows cuff buttons on the white and khaki tunics.

From a lithograph by **Moritz Ruhl, Leipzig.**

**Officer**
wearing the Grey Home Uniform as Parade Dress

**Officer**
wearing the White Tropical Uniform

**Corporal**
wearing the White Tropical Uniform

**Lance Corporal**
wearing the Khaki Field Uniform

the budget of 1895/96 maintained four European officers and twelve NCOs, this number was reduced in 1897 to two officers, a doctor, eight NCOs and only 255 African soldiers. In 1898, a modest reinforcement of the force took place. Governor von Puttkamer did not rest until, in 1901, the *Schutztruppe* finally achieved a standard strength of six field companies, an artillery detachment and a depot company, with a total 101 European and 900 African soldiers. It remained as such until 1905.

Two more companies were formed in 1906, and a further two in 1910, bringing the total to ten companies. By the outbreak of war in 1914, there were twelve companies of the Cameroon *Schutztruppe*.

## Uniforms, Equipment & Weapons of the *Schutztruppe*

### German Officers and NCOs

The uniform for German officers and NCOs of the Cameroon *Schutztruppe* was the same as that of the East African *Schutztruppe*, except that the insignia colour and underlay of the officers' shoulder straps was ponceau red. The piping on the khaki uniform, however, was cornflower blue.

The hat band and edging of the *Südwester* hat were of ponceau red silk ribbon for officers and ponceau red wool for NCOs. The hat band and piping on the upper edge of the cap were of ponceau red cloth. The collar of the officer's *Paletot* coat was of ponceau red cloth both inside and out, and the greatcoat of the NCOs had ponceau red shoulder straps and collar patches.

African Soldier of the *Schutztruppe*, 54 mm figure painted by the author.

The tropical helmet could be white or khaki. Boots and lace-up shoes with leather gaiters were worn.

### African Soldiers in Cameroon

The uniforms of the African soldiers initially consisted of khaki drill jackets in a sailor's cut with red piping and short, knee-length trousers. Later, the uniform followed that of the East African *Askari*, a khaki tunic, but with red edging and braid on the collar. On the cuffs of the tunic was a red band meeting at a point. Four or five plain metal buttons fastened the front of the tunic. The trousers were also made from khaki drill, with wound-cloth puttees and brown leather laced ankle boots. These boots were often not worn

The first Police Soldiers for Cameroon under the leadership of *Leutnant* von Stettin, 1889.

on the march. The Cameroon soldier did not wear the *Tarbusch* with neck shade, but rather a short red rolled Fez with a black tassel. At the front was a small silver flying Imperial eagle badge.

The NCOs' arm chevrons were red with a white background; one chevron for a *Gefreiter*, two for an *Unteroffizier* and three for a *Sergeant*. Musicians wore white swallows' nests with red edging. The African *Feldwebel* was issued with the *Südwester* hat in the style of the German NCOs, with hat band and edging in red.

Equipment was the same as for the *Askari* in German East Africa, except that the soldiers of the three mounted companies had cartridge belts similar to that of the South West African *Schutztruppe*, but with shoulder straps crossing over the breast. Their armament consisted of the *Jägerbüchse M71* rifle, the *Karabiner M88*, the *Karabiner M98* and the corresponding *S71/84* or *kS98* bayonet.

The rolled Fez for African Soldiers in the Cameroon *Schutztruppe*.

Guard Detachment in Garua.

African Soldiers of the Cameroon *Schutztruppe*.

L to R: *Feldwebel* Mboari, twice decorated with the Combatant's Merit Medal, and *Unteroffizier* Ssanga.

The German Flag is raised.

Rifle Cleaning.

Kapelle der Kaiserl. Schutztruppe Kamerun (Duala) Kapellmeister P. Henschel.

Musicians of the Cameroon *Schutztruppe* Band.

Cameroon *Schutztruppe* Officer's *Südwester* Hat.

Cameroon *Schutztruppe* Officer's Cap.

Home Uniform of a *Hauptmann* in the Cameroon *Schutztruppe*.

*Cameroon*  201

Photos on these two pages:
Fotoarchiv Firma Helmut Weitze
Militärische Antiquitäten KG

Cameroon *Polizeitruppe*
Senior NCO's Sword.

*Feldwebel* Mballa of the 7th Field Company in Garua.

Mounted African *Feldwebel*.

*Feldwebel* with his African Other Ranks.

*Unteroffizier* of the *Schutztruppe*.

African Soldiers of the Outpost at Fianga.

An Officer's birthday party.

Cameroon *Schutztruppe* in Soppo, 1912.

Nummer 31.     Seite 1315.

# Beim Gouverneur von Kamerun.

### Vom Bezirksamtmann Dr. Mansfeld. — Hierzu 5 photographische Aufnahmen.

Der Sitz des Gouvernements und zugleich der Wohnsitz des Gouverneurs Dr. Seitz befindet sich in Buea. Der Ort liegt am Südabhang des Kamerunberges, der unter dem Namen Mango=ma=Loba (Götterberg) bekannt ist; der Gipfel des Berges ist 4080 Meter hoch. Man erreicht Buea, indem man von Victoria, dem ersten Hafenplatz für die von Europa kommenden Dampfer, per Bahn in drei= bis vierstündiger Fahrt nach dem Ort Soppo (750 Meter hoch) fährt und von dort zu Fuß oder per Pferd oder Wagen nach dem 1000 Meter über dem Meeresspiegel liegenden Buea weitergeht. Der Urwald beginnt direkt am Meer in Victoria und reicht weit über Buea hinaus bis zu 2000 Meter Höhe; dort wird er durch Grasland abgelöst; auf dem Gipfel liegt hie und da, besonders im Monat April, auf Stunden, manchmal auf Tage eine leichte Schneedecke. Das längs der Küste sich hinziehende flache Vorland schmilzt in Victoria zu einem schmalen Streifen zusammen, der Berg steigt fast unmittelbar aus dem Meer empor; daher der wundervolle Anblick bei der Einfahrt in den Hafen vom Dampfer aus. Ich stehe nicht an, den Hafen von Victoria als den landschaftlich schönsten unter allen Häfen, die ich in Südamerika, Indien, Ostasien und Afrika von Marokko

Von links: Adjutant v. Puttkamer, Frau Seitz, der Gouverneur, Frl. v. Cleve.
**Beim Tee im Garten. Oberes Bild: Vor dem Ausritt.**

Magazine article about Dr. Theodor Seitz, Governor of Cameroon, 1912.

Illustration of a *Schutztruppe* Caravan crossing a Rope Bridge, 1903.

## Organisation of the *Schutztruppe* in 1914

In April 1914, *Major* Carl Zimmermann took command of the Cameroonian *Schutztruppe*. *Oberstleutnant* Paul von Lettow-Vorbeck had first been considered for the command, but instead was appointed to the *Schutztruppe* in German East Africa.

At that point Zimmerman commanded a *Schutztruppe* consisting of twelve companies with a total of 185 German officers, doctors, NCOs, and military officials, as well as 1,550 African soldiers, including recruits. That included three mounted units and an additional artillery detachment. The deployments of the *Schutztruppe* were arranged as follows:

**Staff HQ:** Soppo

- **1st (Depot) Company** with six machine guns. Artillery Detachment (with four old 9 cm *M73/91* field guns) and garrison troop at Duala.
- **2nd Company** with four machine guns in Bamenda, with outposts at Kentu & Wum.
- **3rd Company** with three machine guns at Mora, with an outpost at Kusseri.
- **4th (Expedition) Company** with four machine guns at Soppo.
- **5th Company** with three machine guns at Buar, with an outpost at Carnot.
- **6th Company** with three machine guns and a 3.7 cm gun at Mbaiki, with an outpost at Nola.
- **7th Company** with three machine guns and two 6 cm guns at Garua, with outposts at Marua, Mubi, Lere and Nassarau.
- **8th Company** with three machine guns and a 6 cm gun at Ngaundere.
- **9th Company** with three machine guns at Dume, with outposts at Baturi and Buma.
- **10th Company** with two machine guns and a 3.7 cm gun at Ojem, with an outpost at Mimwoul.
- **11th Company** with three machine guns and a 3.7 cm gun at Akoafim, with outposts at Ngarabinsam and Minkebe.
- **12th Company** with three machine guns at Bumo, with outposts at Kodjala and Fianga.

*Polizeitruppe* training in Buea.

German Officers with African Soldiers and their wives.

## Cameroon *Polizeitruppe*

In order to carry out the tasks of civil administration, a separate *Polizeitruppe* not subordinate to the *Schutztruppe* was established. Its strength gradually grew to about 550 men. The uniforms, equipment and weapons matched those of the *Schutztruppe* almost entirely. The *Polizeitruppe* was distributed in units large and small to the various administrative districts. These units were commanded by German Police NCOs (*Polizeimeistern*). There was also a depot company in Duala where the *Polizeitruppe* received their military training.

Mounted African *Unteroffizier*.

Cameroon *Schutztruppe* with a 3.7 cm Quick-Firing Gun.

Cameroon *Schutztruppe C73* Heavy Field Gun Battery.

*Schutztruppe* with a 3.7 cm Quick-Firing Gun in Cameroon.

Above: *Schutztruppe* Reinforcements departing for Cameroon, November 1904.   Below: Defensive Position, 1915.

## The First World War in Cameroon

On 2 August 1914, a state of emergency was declared in the colony, followed by British and French invasions. The *Schutztruppe* called upon German and African reservists and the *Polizeitruppe* until it was more than twice its previous strength, and fought a series of delaying actions which lasted eighteen months. In January 1916, desperately low on ammunition, the bulk of the *Schutztruppe* marched into the neighbouring neutral Spanish colony of Rio Muni to be interned. The last bastion of German resistance in Cameroon surrendered at Mora on 18 February 1916.

With the Spanish authorities in Muni unable to accommodate this influx, the *Schutztruppe* were moved to Spanish-owned Fernando Po, an island 20 miles off Duala, the capital of Cameroon. This lingering threat concerned the Allies, whose troops were needed elsewhere. They suspected the Spanish of selling weapons to the *Schutztruppe*, and protested this breach of neutrality. To defuse the threat, the Spanish asked the *Schutztruppe* to give up some of their weapons including their machine guns, and deported some senior officers to Spain. Lax supervision allowed several officers to return to Germany and rejoin the war.

Though their strength was diminished, a German force remained on Fernando Po throughout the war, ready to re-take the colony in the event of a German victory in Europe, which of course never came.

Before the war ended, Great Britain and France concluded a secret treaty in which Cameroon was divided between the two countries, with France having the larger administrative territory.

Cameroon *Schutztruppe* Artillery Team.

Photo: *Sueddeutsche Zeitung* / Alamy Stock Photo

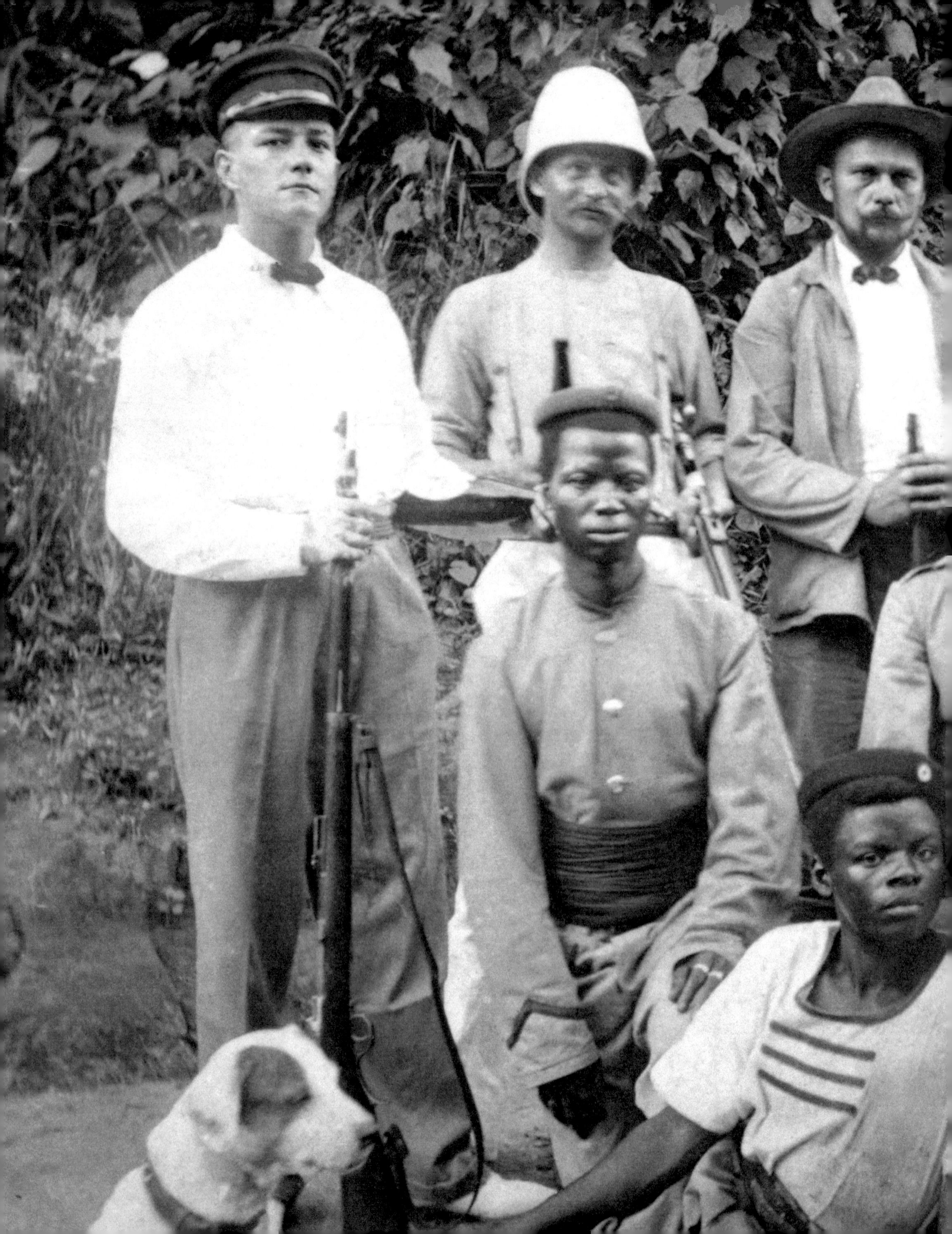

Off-Duty members of the Cameroon *Schutztruppe.*

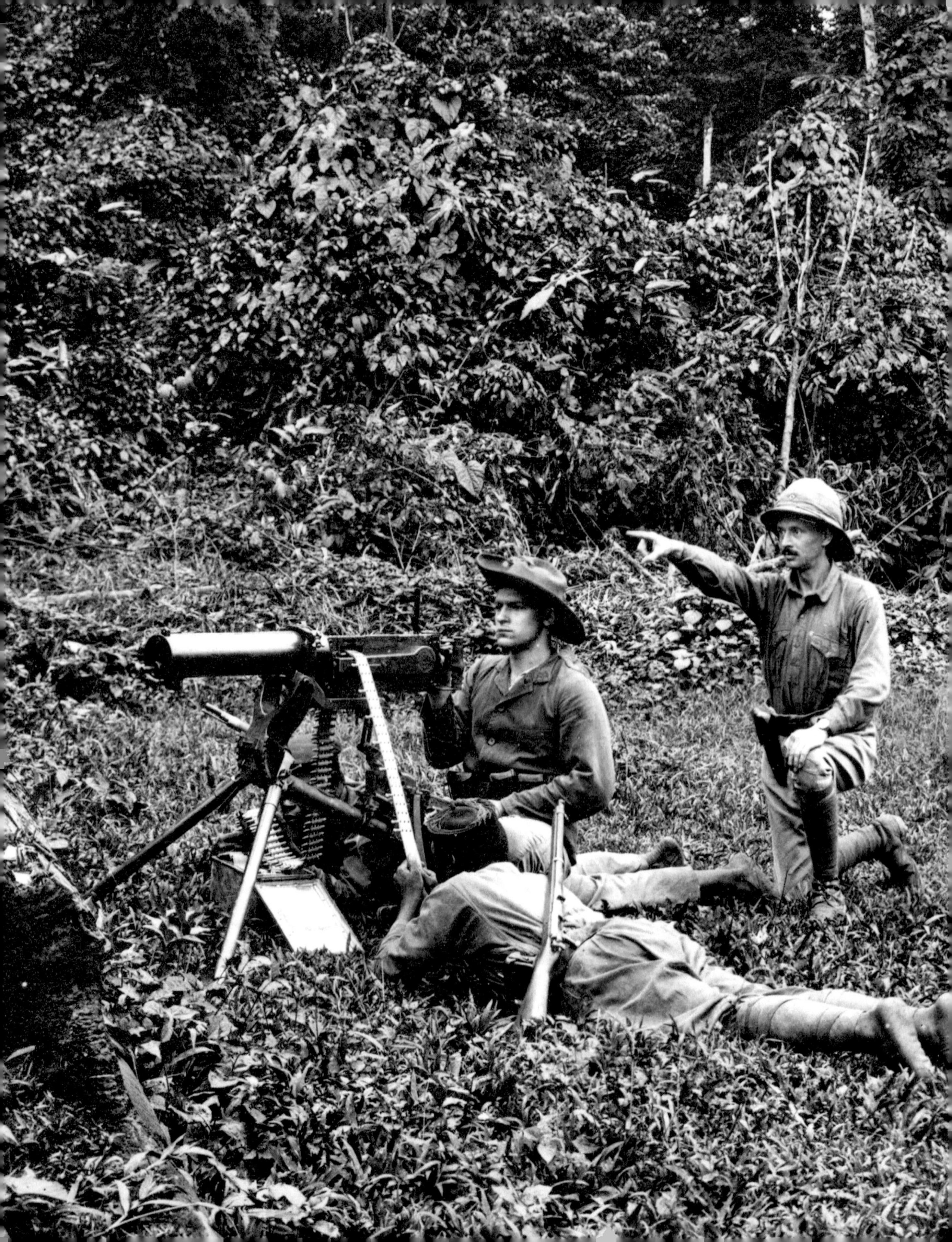

Machine Gun Practice in Cameroon.

Cameroon *Schutztruppe* with a Spanish Soldier in Bata, Spanish Muni, early 1916. Note the *MG08* Machine Gun and large numbers of Rifles being surrendered by the Germans. Many of these Rifles were previously captured from the French. The Cameroon *Schutztruppe* crossed the border to Muni in January 1916 with 95 German Officers, 450 German NCOs and Other Ranks, and roughly 5,000 African Soldiers.

Accompanying them were 400 German Civilians and around 40,000 African Civilians including Porters and their families but also large numbers of the Beti people who had remained loyal to the Germans.

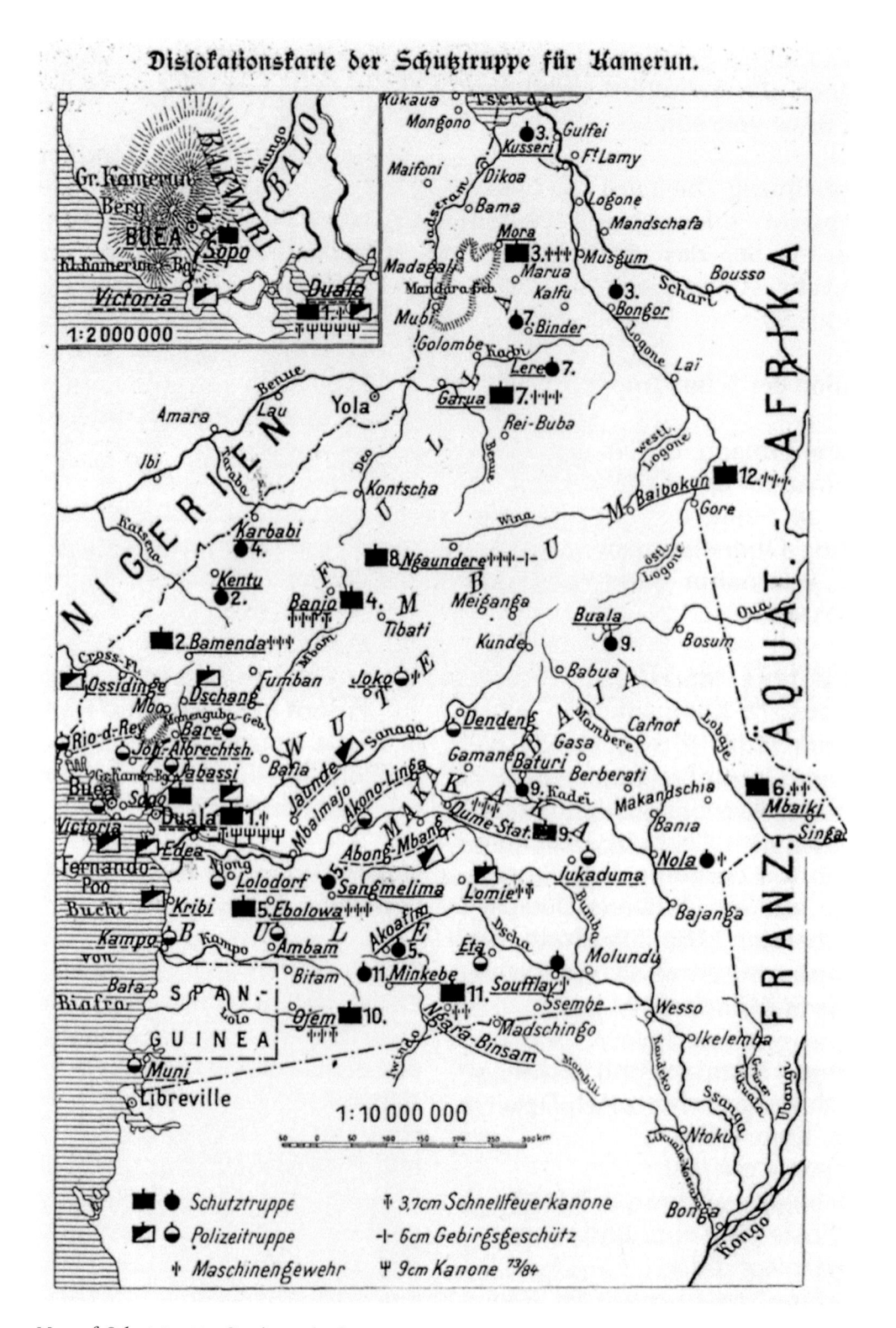

Map of *Schutztruppe* Garrisons in Cameroon.

## Cameroon *Schutztruppe* March

Words by A. von Engelbrechten,
music by Willy Weide

Whether we find ourselves happy at home again,
Or if we are soon resting in this earth,
It is alright, because here we rest in German ground,
High, three times high! Hurrah, my Cameroon!

*Ob wir daheim uns froh einst wiederfinden,*
*Und ob wir bald in dieser Erde ruhn ,*
*Stimmt ein, denn hier ruhn wir in deutschen Gründen,*
*Hoch, dreimal Hoch, Hurra, mein Kamerun!*

Cameroon *Schutztruppe* Soldiers at a Guard Post.

# Map: Togo and surrounding regions

- Sokoto
- Say
- (Franz.) Boti
- Gandu
- Wagadugu
- GANDU
- Pamma
- Gomba
- Gambaga
- BORGU
- Susanne Mangu
- Jendi
- Bussang
- Tewande
- Bassari
- Tegv
- NUPE
- Paratau
- Rabba
- Bismarckburg
- Sarakin
- Salaga
- Eggar
- Kratschi
- Franz.
- Ibadan
- Atak
- Dahome
- Lokodj
- Kpando
- Oparae
- JORUBA
- Misahöhe
- Agome
- Abeokuta
- ASCHANTI
- Ahmedyow
- umassi
- Ho Togo
- Lagos
- abubu
- Brit.
- BENIN
- Abetifi
- Assabo
- Akropong
- Volta
- Cape Coast Castle
- Christiansborg
- Benin Bay
- Gold-Küste
- Sklaven-K.
- MEERB. v. GUINEA

# Part IV
# **Togo**
*(Togo)*

Map of Togo and its Hinterland, 1906.

## Commissioners and Governors of Togo

1884 – 1885  Julius Freiherr von Soden (with the title of *Oberkommissar für Togoland*)

1885 – 1889  Ernst Falkenthal (with the title of *Regierungsassessor, Kommissar für Togoland*)

1889 – 1895  Jesko von Puttkamer (with the title of *Landeshauptmann* since 1893)

1895 – 1902  August Köhler (with the title of *Gouverneur* since 1893)

1902 – 1903  Waldemar Horn

1904 – 1910  Johann Nepomuk Graf Zech auf Neuhofen

1910 – 1912  Edmund Brückner

1912 – 1914  Herzog Adolf Friedrich zu Mecklenburg

1914  *Major* Hans-Georg von Döring, with the title of Deputy Governor (*Stellvertretener Gouverneur*), as Herzog Adolf Friedrich was on leave in Germany when the First World War broke out

## Formation of the *Polizeitruppe*

In 1885, the first German Commissioner in Togo, Ernst Falkenthal, recruited a small force of eight Hausa soldiers to perform police duties at the seat of government. This small force naturally proved to be far too weak to accomplish extensive tasks. The Commissioner therefore requested the establishment of a *Polizeitruppe* later that same year. An *AKO* of 30 October 1885 set the provisional strength at one German sergeant and ten Hausa soldiers. The history of the *Polizeitruppe* in Togo began that day.

L to R: Seals of the Governor of Togo and *Landeshauptmann* of Togo.

The tasks of this force were initially limited to local security service. If further military forces were necessary it was intended that landing parties from German cruisers could be deployed. In the end, expansion of the *Polizeitruppe* proved necessary and so their numbers rose from 35 in 1886 to around 50 in 1888 and then 92 by 1890. In October 1891, the force was reduced to 30 men for budgetary reasons but due to the defiance of various tribal chiefs such low force levels soon proved completely insufficient. Governor von Puttkamer enlisted another 45 men in the course of 1893 and in 1894 received permission from the Imperial chancellery to completely reorganise the *Polizeitruppe*.

Recruitment was initially confined to the militant Hausa tribes but later included the Grussi and Mossi tribes, the Dahomey and Wey people, as well as the more indigenous people of the colony of Togo such as the Dakomba, Konkomba, Tchokossi, Losso and Kabure.

The officers of the *Polizeitruppe* did not leave the army but were seconded by Imperial decree to serve the Colonial Office. Thus they remained under disciplinary control of their unit. The German *Polizeitruppe* NCOs, on the other hand, had left military service and then transferred to the civil service of the colony, similar to members of the Royal Guard in Berlin.

The *Polizeitruppe* in Togo were employed solely for military tasks. For the actual police work, local administrations employed African policemen subordinate to that administration and not belonging to the *Polizeitruppe* in any way.

The difference between the *Polizeitruppe* of Togo and the *Schutztruppe* of Cameroon, German East Africa and German South-West Africa was in its command structure, since the Togo *Polizeitruppe* had no commander but was directly subordinate to the civil government or colonial governor.

African Chieftan in Togo swearing Allegiance to the German Flag.

# Uniforms, Equipment & Weapons of the *Polizeitruppe*

The uniform for German officers and NCOs of the Togo *Polizeitruppe* was the same as for the Cameroon *Schutztruppe*, except that the insignia colour of the backing of the officer's shoulder strap was yellow. This also applied to the hatband and the brim-edging of the hat. Those of officers were made of yellow silk ribbon, and those of NCOs made of wool ribbon. The hatband and upper rim piping on the cap were also of yellow cloth.

In addition to German tropical helmets, helmets manufactured by French and English firms were also worn.

The uniform of the African soldiers of the Togo *Polizeitruppe* was subject to many modifications over the years. The original uniform of blue serge was replaced by a white drill uniform with a red fez and a red body sash. In addition the soldiers carried the *Karabiner M71* and a bayonet.

A decree from 1887 then designated the uniform as being a jacket and baggy trousers of blue wool with red edging, a white field cap with white peak, and a red sash. Equipment consisted of a belt and cartridge box, bread bag, cookware, and canteen with felt cover on a strap. The bayonet was the *Pionier-Faschinenmesser M71*. In 1888 the newly hired recruits received the Gewehr *M71* rifle, and gradually the *Karabiner M71* carbine lacking a bayonet fixture were decommissioned.

Governor's Residence in Lome.

Reception in Lome for Duke Adolf Friedrich of Mecklenburg.

Bugle used by African Troops in German Service.

Togo *Polizeitruppe*.

After 1890 the white peaked cap was replaced by a blue peakless infantry cap, which in turn was soon replaced by a turban-like red rolled fez with flying silvered Imperial eagle and a blue/black tassel. The fez remained in use until 1910, at which point khaki peaked caps with ponceau red edging and a German cockade were introduced.

The uniform changed again on the occasion of the reorganisation of the troop in 1894, at which point the blue woollen uniforms were replaced by those of khaki with a low stand-up collar and bearing red collar insignia (*Litze*). Shoulder straps on this uniform were initially red but later khaki. Knee-length trousers were of the same colour, with no piping. Shoes and puttees were not worn by the African soldiers.

The *Pionier-Faschinenmesser M71* bayonet was replaced by the *Hirschfänger M71* with a natural leather belt and two front cartridge pouches. The equipment was completed by African-style water bottles, knapsacks, and horse blankets used as sleeping covers, which were rolled up and worn over the existing equipment across the chest. Starting in 1907 soldiers also carried a tent section, which was rolled with the blanket around the backpack, as well as hatchets and picks like the infantry. In 1909 they were also equipped with a broad machete for clearing pathways in dense bush. At the same time, the *Seitengewehr M71/84* bayonet was introduced to replace the *Hirschfänger M71*. The *Jägerbüchse M71* rifle also gradually replaced the *Gewehr M71*. The *Jägerbüchse* light infantry rifle was fitted with a cartridge ejector to increase the rate of fire. Its simple and durable construction made it very suitable for bush warfare against natives, as both the rifle and its cartridges were able to withstand the ravages of the tropical climate. However, a modern enemy with a magazine rifle and smokeless powder put the *Jägerbüchse* at a disadvantage, as was proven during the First World War in Togo, Cameroon, and German East Africa.

Up until 1896 rank was shown as bars worn parallel to the gold cuff lace. From then on, a Lance Corporal (*Gefreiter*) wore one red cloth chevron above the cuff braid, a Corporal (*Unteroffizier*) had two red chevrons, a Sergeant (*Sergeant*) had three, and the Sergeant Major (*Feldwebel*) four, arranged one above the other.

*Polizeitruppe* in Lome.

"Decree concerning the clothing, equipment and weaponry of the *Polizeitruppe*, local police, border guards and chieftain policemen, is hereby fixed as follows:

**A. *Polizeitruppe***

a. Clothing and Equipment:

Tunic of the pattern already introduced in the colony, with white metal buttons. Trousers of the pattern already introduced in the colony. Navy shirt. Cap of the pattern already introduced in the colony with German cockade. Red cloth sash.

Wool rug with three straps. Backpack with carrying frame of the pattern already introduced in the colony. Tent equipment consisting of tarpaulin, three tent poles, three tent pegs and one tent line. Cookware with attachment straps. Belt with plain clasp and bayonet frog; two front-mounted cartridge pouches. Bread bag with strap. Aluminum canteen with cover, belt attachment and cork with leather straps. Entrenching tools: spades and picks as needed; any soldier who does not carry a pick or spade has a machete of the pattern already used in the colonies.

Buglers also carry the Prussian infantry bugle, and drummers their marching drum and special pad.

*Vizefeldwebel*: instead of the short trousers, long khaki trousers and leather lace-up boots; on garrison duty, instead of the other ranks' belt and bayonet frog, a senior NCO's buckle and sabre with scabbard and with, for the *Feldwebel* and NCOs, a sword knot. All NCOs also carry the rifle and its strap.

b. Weaponry:

*Jägerbüchse M71* or *Gewehr M71* with muzzle cap, rifle strap and *M71* or *M71/84* cartridges. Machine gun crews carry the *Karabiner M71*.

*S71/84* Bayonet. *Vizefeldwebel*: Infantry sword. In the field, the *Jägerbüchse M71* and *Bayonet 71/84*.

**B. Local Police (*Ortspolizei*)**

Similar to the *Polzeitruppe* but with the following differences:

Clothing and Equipment:

Tunic: a red cloth P sewn on the left sleeve, yellow metal buttons; black sash. The backpack, tent equipment and cookware are dispensed with.

Weaponry:
Instead of the *Jägerbüchse 71*, they are armed with the *Karabiner M71*.

**C. Border Guards (*Grenzwächter*)**

Similar to the *Polizeitruppe* but with the following differences:

Clothing and Equipment:

Tunic: a red cloth Z sewn on the left sleeve, yellow metal buttons; black sash. The backpack and cookware are dispensed with.

Weaponry:

Instead of the *Jägerbüchse 71*, they are armed with the *Karabiner M71*.

**D. Chieftain Policeman (*Häuptlingspolizisten*)**

Only the tunic, trousers, shirt and cap like the *Polizeitruppe*; but with yellow metal buttons and blue sash.

Regulations in Common:

The ranks wear the following red cloth insignia sewn on both sleeves of the tunic:

*Gefreiter*: one chevron

*Unteroffizier*: two chevrons

*Sergeant*: three chevrons

*Vize-Feldwebel*: four chevrons

Musicians wear swallow's nests made of red and white cloth on both shoulders. Marksmanship insignia according to the pattern already introduced in the colony. Older items that deviate from the rules given here are still permitted."

*Editor's Note: A further copy of that order with additional references was sent from the Imperial Government of Togo at Lome on 12 April 1913, as follows:*

"The modification of the uniform of the *Polizeitruppe* of the colony, which was decided at the last district meeting, will come into effect on 1 January 1914. The old uniforms can be worn until the new uniforms are ready to take their place. The change in the uniform consists mainly of the following: introduction of the long cut jacket like waist tunic without skirts, discontinuation of the sash and the introduction of yellow insignia.

The new uniform is more in line with the requirements of field regularity than the previous one. Yellow badges are introduced because yellow is the basic colour introduced for the headgear, etc., of the officers deployed to Togo. The Governor's tunic is piped yellow, and also the shoulder strap base is yellow in colour, so yellow is the basic colour for the Togo uniform.

Description of the new uniform:

1) The headgear is the corduroy field cap with peak, in the style worn by the South-West Africa *Schutztruppe*.

At the moment the troops in Lome are testing the field cap; the final decision will rest on their results. There is a thin yellow piping around the rim of the cap and the same goes around the lower and upper edge of the cap band.

2) The tunic is like the home *Waffenrock* tunic, cut down to size and without the hip skirting attached. It is closed by five white metal buttons. The tunic is slit open on the lower rear at the right and left.

On the shoulders are buttoned shoulder straps. On the front left and right of the collar (which is closed by one hook), is a yellow lapel 4 cm wide and 5 cm long.

Around the forearms there is a yellow band instead of the previous red. All ranks from *Gefreiter* up to have a yellow braid running around the neck of the collar, so that from now on the rank will be also recognisable from behind. NCO rank insignia is worn only on the left upper arm in the new uniform. Number of chevrons as before.

3) The cut of the trousers stays the same as before. Held together at the waist with a sewn-in band.

4) Puttees and sandals will be introduced as footwear. Attempts at this are still being made.

The Governor, Lome, Togo"

*Editor's Note: There followed an additional note from the Imperial Government of Togo at Lome, dated 30 April 1913.*

"On the topic of Request of 24 April 1913.

The Governor's decree of 12 April 1913 No. 2547/13 on the change of uniforms relates only to the members of the *Polizeitruppe* and not to the local police.

The Governor."

*Editor's Note: This is a transcript of the Governor's correspondence concerning the Uniforms and Equipment of the Togo Polizeitruppe in 1913.*

Togo *Polizeitruppe*, ca. 1885 – 1887.

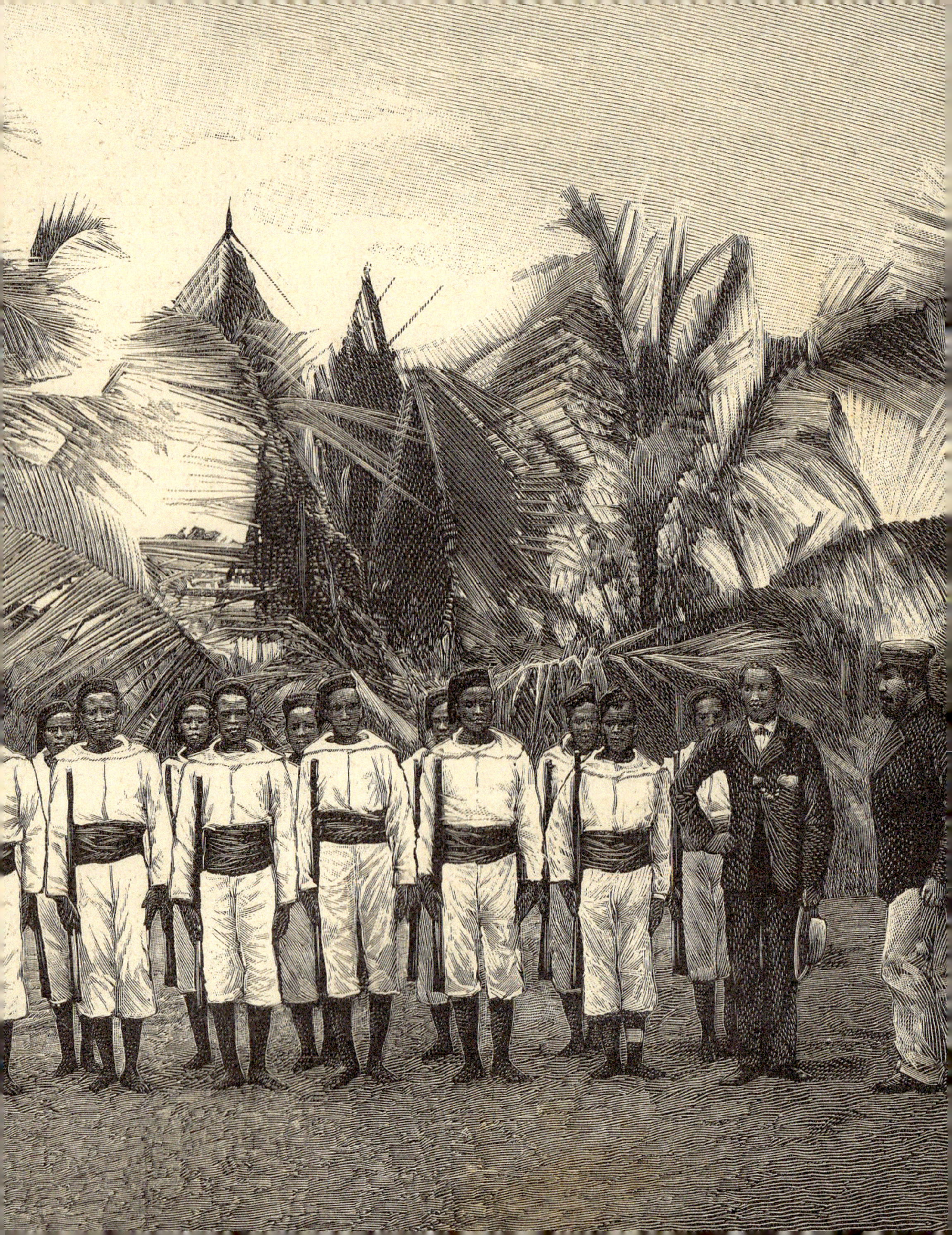

Togo — Eingeborene Soldaten

Togo *Polizeitruppe.*

Togo *Polizeitruppe* in Training. Note the NCOs wear the *Polizeitruppe* Field Cap authorized in 1913.

Photo: *Süddeutsche Zeitung* / Alamy Stock Photo

New Recruits for the Togo *Polizeitruppe*.

Photo: Süddeutsche Zeitung / Alamy Stock Photo

Sailors of the Imperial Navy in Togo with African *Polizeitruppe* and villagers.

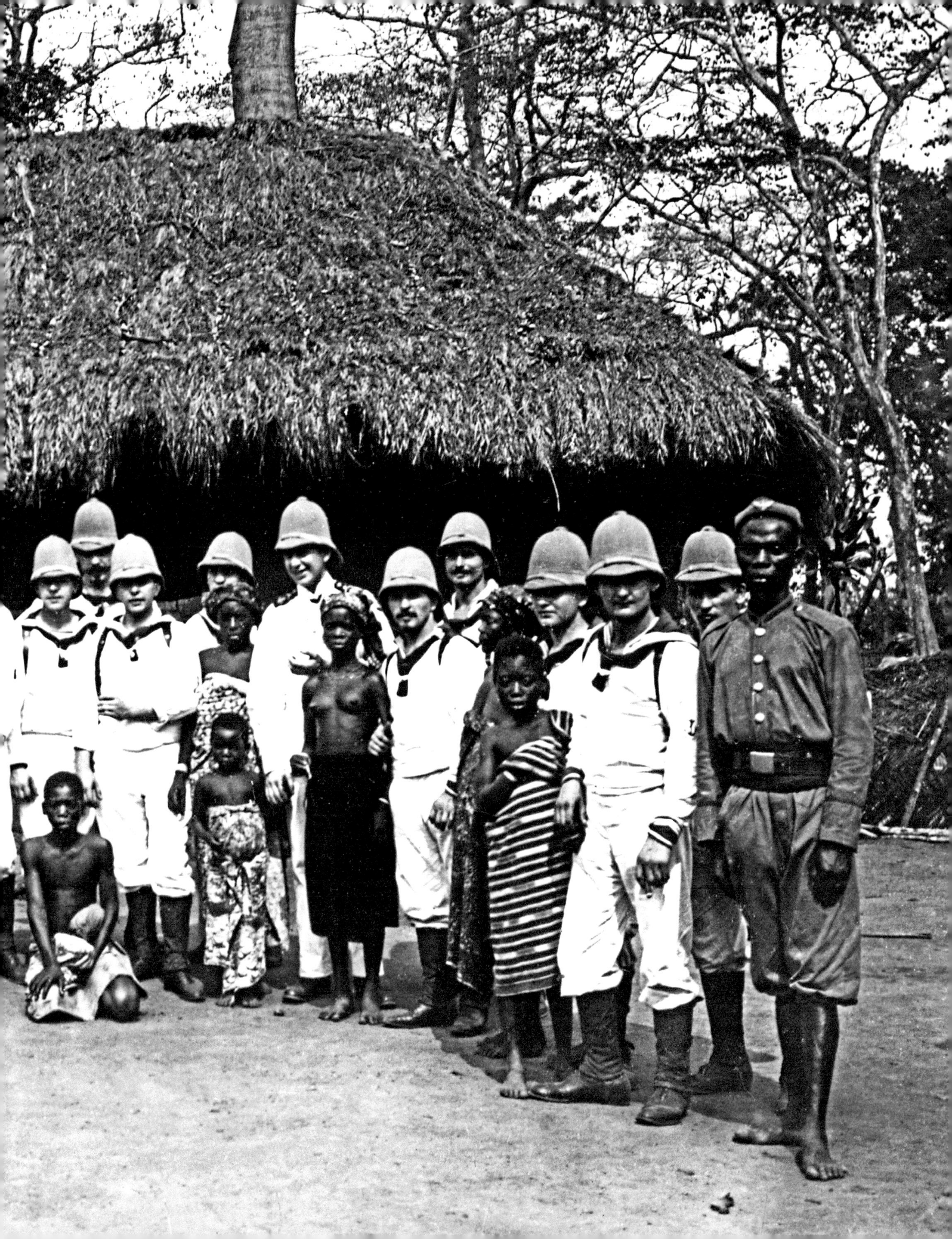

## Organisation of the *Polizeitruppe* in 1914

The *Polizeitruppe* now consisted of about 530 men, of which around 150 were stationed in the capital Lome, while the rest were distributed in units of 30 to 60 men at various stations around the interior. The troop was subordinate to two **Police Inspectors** (*Polizeiinspektoren*), a **Captain** *(Hauptmann)* and a **First Lieutenant** (*Oberleutnant*). These two men belonged to the home army, had been seconded to the Colonial Office, and then deployed to the colony. There were no real German NCOs, only a number of **Police Officers** (*Polizeimeister*) who were civil servants. Of course, only former NCOs of the army served the *Polizeitruppe* as drill masters.

Recruitment now took place throughout the colony itself, mainly in the northern districts. Incidentally, the company in Lome was not the depot company of the *Polizeitruppe* from which each of the various police groups were supplied with soldiers. Rather, each district found its own recruits.

## The First World War in Togo

On 5 August 1914, the colony's Deputy Governor, *Major* Hans-Georg von Döring, sought assurances of neutrality from Britain and France. As the British answer was evasive and the French did not respond at all, a state of war was declared in the colony.

After skirmishes against invading British and French forces, on the night of the 24/25 August the radio station at Kamina was destroyed. Following negotiations with the British, at 8 am on the morning of 27 August, Togo was handed over to the Allies. In 1916, it was divided between Britain and France under the terms of a secret treaty.

Photo of a Police Station in Togo, from *Deutsche Kolonial Zeitung*.

## Togo Poetry Verses

The following anonymous verses were found by the Post and Telegraph Inspector Dr. Max Roscher, in the interior of Togo on a reconnaissance mission during the defence of the radio station at Kamina in 1914.

| | |
|---|---|
| I look at the dark blue | *Blick ich auf die dunkelblauen* |
| Atakpame Mountains there, | *Berge Atakpames hin,* |
| Taking me back to the old grey | *Kommen mir die altersgrauen* |
| Castles of Germany in my mind. | *Burgen Deutschlands in den Sinn.* |
| I think of the happy hours, | *Und ich denke froher Stunden,* |
| Without my heart complaining. | *Ohne daß mein Herz beschwert,* |
| Here I have found, | *Hab' ich doch hier gefunden,* |
| What a German man is worth. | *Was des deutschen Mannes wert.* |
| | |
| I see German pennants, | *Deutsche Wimpel seh' ich wallen,* |
| On the ocean free and happy, | *Auf dem Weltmeer frei und froh,* |
| I hear German lutes sounding | *Deutsche Laute hör' ich schallen.* |
| From Mangu to Anecho. | *Von Mangu bis Anecho.* |
| I also live in this distant land, | *Leb' ich auch im fernen Lande.* |
| Yet I feel close to home | *Weht doch an Lomes Strande,* |
| As your banner, *Germania* | *Dein Panier, Germania."* |
| Flies on Lome beach. | |

On the 27th of January, 1912, the 53rd Birthday of Kaiser Wilhelm II, *Schutztruppe* Soldiers, Government Officials, Clergy and Civilians gather in Windhoek, German South-West Africa to witness the unveiling of Sculptor Adolf Knerle's bronze *Reiterdenkmal* Statue.

Colorization by Reinaldo Elias.
Photo: *Süddeutsche Zeitung* / Alamy Stock Photo.

# Part V
# Medals, Campaign Clasps, and Recognitions of Service

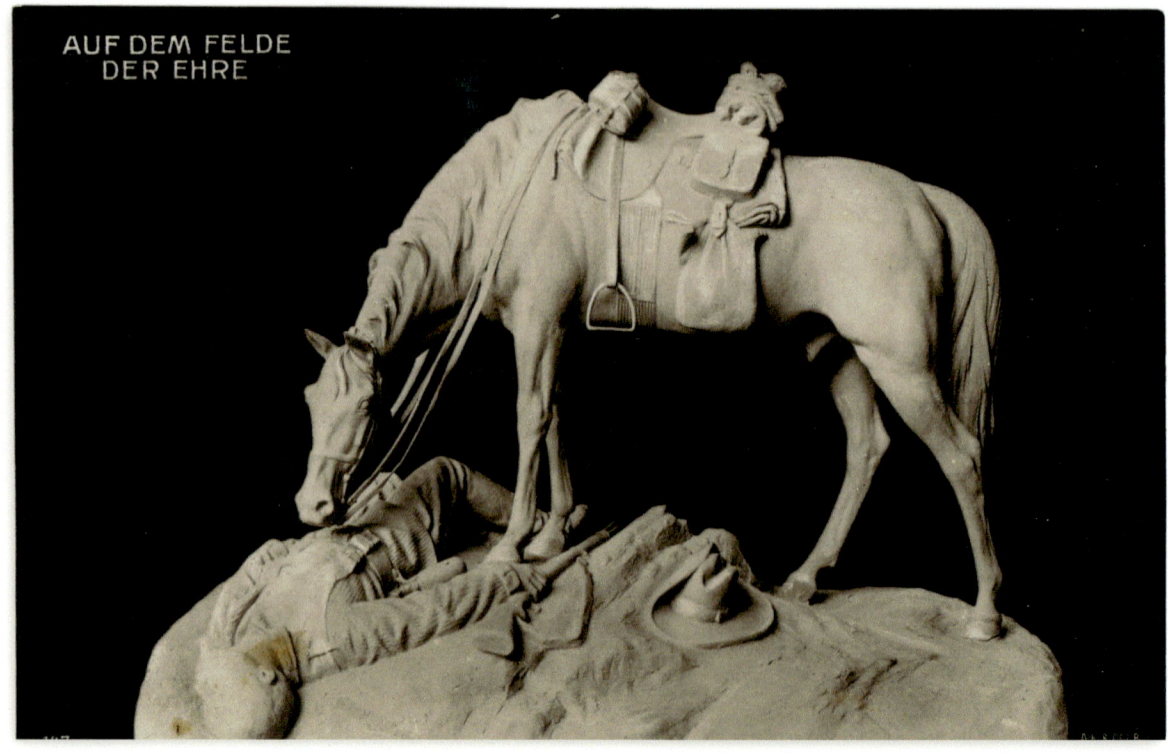

Postcard showing the South-West Africa Medal (*Südwestafrika-Denkmünze*), for both Combatants (left) and Non-Combatants (right). Seen above the Medals are Campaign Clasps for "WATERBERG" and "HEREROLAND".

Below: Postcard of fallen *Schutztruppe* Rider, with inscription "On the Field of Honor"

## South-West Africa Medal (*Südwestafrika-Denkmünze*)

The South-West Africa Medal was issued by Kaiser Wilhelm II beginning on 19 March 1907. It was awarded in bronze for all German forces involved in the suppression of the Herero and Nama uprisings in South-West Africa from 1904 to 1908, as well as for personnel involved in the care of the sick and wounded.

It was awarded in steel to people who had been extraordinarily involved in the preparations for the deployment of the armed forces, as well as to members of the crews of ships of German shipping companies which were chartered to carry troops and war supplies to and from South-West Africa.

The exact number of awards cannot be reliably determined. It is estimated that there were around 20,000 bronze awards and 6,000 awards of the steel medal.

Award Certificate for the South-West Africa Medal (*Südwestafrika-Denkmünze*), 1907.

South-West Africa Medal (*Südwestafrika-Denkmünze*), and Miniatures, obverse. Top row, from left to right: Combatants, Non-Combatants, and gilded for Combatants. Bottom row: Miniatures for Combatants and Non-Combatants.

**Front:** The obverse showed the head of Germania with a winged helmet and armor and the text '*SÜDWEST-AFRIKA 1904 – 1906.*' On the arm of Germania was the name of the designer: 'SCHULTZ'.

**Back:** In the center were the gothic letters "W II", surmounted by the Imperial crown and flying ribbons. The medal for combatants bore the inscription "*DEN SIEGREICHEN STREITERN*" (To the Victorious Fighters) above crossed swords.

The medal for non-combatants had the gothic 'W II' topped by the Imperial crown, but below it was a laurel branch instead of the crossed swords. The inscription read: '*VERDIENST UM DIE EXPEDITION*' (Merit for the Expedition).

The ribbon of the medal was 35 mm wide (sometimes 31 mm) with 5 mm-wide black and white side margins and white sewn edges. The remaining middle part of the ribbon was horizontally striped in red and white. The coloured stripes were each 1 mm wide.

The bronze version could also be privately gilded and made as half-size (*Prinzengröße*n) or miniatures.

South-West Africa Medal (*Südwestafrika-Denkmünze*) and Miniatures, reverse.

## South-West Africa Medal Campaign Clasps

For the South-West Africa Medal, sixteen campaign clasps were awarded, recalling either a single battle or several battles over a longer period. The clasps were titled:

HEREROLAND, OMARURU, ONGANJIRA, WATERBERG, OMAHEKE, GROSSNAMALAND, FAHLGRAS, TOASIS, KARAS-BERGE, GROSS-NABAS, AUOB, NURUDAS, NOSSOB, ORANJE, KALAHARI 1907, KALAHARI 1908.

The clasp for Kalahari 1907 was exclusively awarded to British servicemen for their role in helping suppress the Herero and Nama uprisings, namely for killing the rebel leader Jacobus Morenga in battle on September 19th, 1907. In January 1909, members of the British Cape Colonial Forces, notably the Cape Mounted Riflemen and the Cape Mounted Police, received a total of 105 bronze medals and 92 campaign clasps bearing the inscription 'KALAHARI 1907'.

## Colonial Medal *(Kolonialdenkmünze)*

The Colonial Medal was instituted on 13 June 1912 by Kaiser Wilhelm II for all participants in military operations in the colonies. Participants in the military campaign in East Asia and the suppression of the uprising in South-West Africa from 1904-08 were excluded, as commemorative medals had already been awarded for those actions.

The medal could be awarded to all members of the Imperial Army and Navy, the *Schutztruppe* and *Polizeitruppe* in the colonies, and others who participated in operations.

The medal was designed by Professor Rudolf Marschall in Vienna.

**Front:** The bust of the Kaiser in the uniform of a Prussian field marshal. In the lower right quarter was the monogram '*W II*' with the Imperial crown above.

**Back:** The Imperial crown with the inscription '*DEN TAPFEREN STREITERN FÜR DEUTSCHLANDS EHRE*' (The Brave Fighters for Germany's Honour) above and below the crown. This was surrounded by a laurel wreath on the left and an oak branch on the right.

The commemorative coin was made of bronze. Again, there were miniatures and gilded versions for medal bars.

It was minted in two versions:
a) German personnel:
   width 32.6 – 33 mm and height 36 mm.
b) African personnel:
   width 28 mm and height 31 mm.

Colonial Medal (*Kolonial-Denkmünze*) and Miniature, obverse and reverse.

# Colonial Medal Campaign Clasps

For the period 1884 – 1914, a total of 89 campaign clasps were established for 273 military operations. The one-line inscription on the clasp had the name of the colony and the year of the campaign, as follows:

### DEUTSCH-OSTAFRIKA

1888/89, 1889/1890, 1889/91, 1892, 1893, 1894, 1895, 1896, 1897, 1897/98, 1898, 1899, 1900, 1901, 1902, 1903, 1905/07, 1911, 1912.

### DEUTSCH-SÜDWESTAFRIKA

1893/95, 1896, 1897, 1897/98, 1901, 1903/04.

### KAMERUN

1884, 1886/91, 1889, 1890, 1891, 1891/94, 1893, 1895/96, 1897, 1898, 1898/99, 1899, 1899/1900, 1900, 1900/01, 1901, 1901/02, 1902, 1902/03, 1903, 1904, 1904/05, 1905/06, 1905/07, 1906, 1907/08, 1908/09, 1911, 1912.

### TOGO

1894/95, 1895, 1896, 1896/97, 1897, 1897/98, 1898, 1898/99, 1899, 1900, 1900/01, 1901, 1902, 1903.

Other clasps for colonial campaigns outside of Africa included those for NEU-GUINEA, SAMOA and VENEZUELA.

Award Certificate for the Colonial Medal (*Kolonialdenkmünze*), 1913.

# Combatant's Merit Medal (*Kriegerverdienstmedaille*)

In 1892 it was decided that the Prussian *Kriegerverdienstmedaille* could be awarded as an Imperial medal to *Askari* in East Africa. From 25 March 1893, it could also be awarded to non-European soldiers in the *Schutztruppe* and *Polizeitruppe* of all colonies.

- 1st Class in gold, 1892 – 1919
- 1st Class in silver for African officers
- 2nd Class in gold, 1892 – 1919
- 2nd Class in silver for African NCOs and other ranks

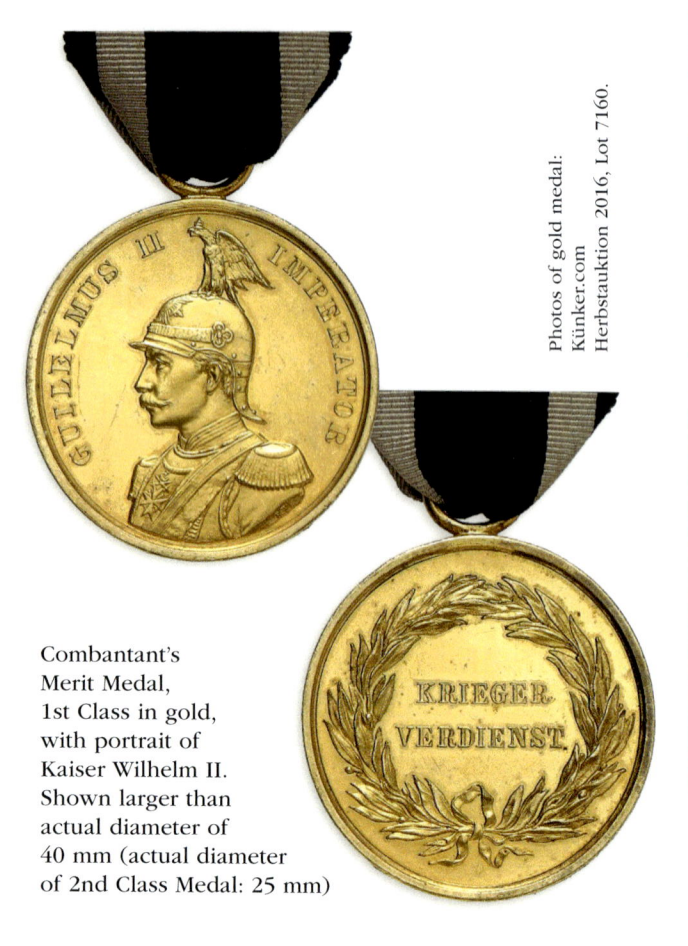

Combatant's Merit Medal, 1st Class in gold, with portrait of Kaiser Wilhelm II. Shown larger than actual diameter of 40 mm (actual diameter of 2nd Class Medal: 25 mm)

Photos of gold medal: Künker.com Herbstauktion 2016, Lot 7160.

Combatant's Merit Medal (*Kriegsverdienstmedaille*), 2nd Class in silver for African Other Ranks, obverse (top) and reverse (bottom).

*Medals, Campaign Clasps, and Recognitions of Service* 251

Photos: Harry Fakner

Left: An African *Feldwebel* of the Cameroon *Schutztruppe* wearing several Medals, including the Prussian Crown Medal and two Combatant's Merit Medals for African Soldiers. Right: An African Soldier of the Cameroon *Schutztruppe* with two Combatant's Merit Medals. Next to him is a German *Schutztruppe* NCO and an African woman, most likely the soldier's wife.

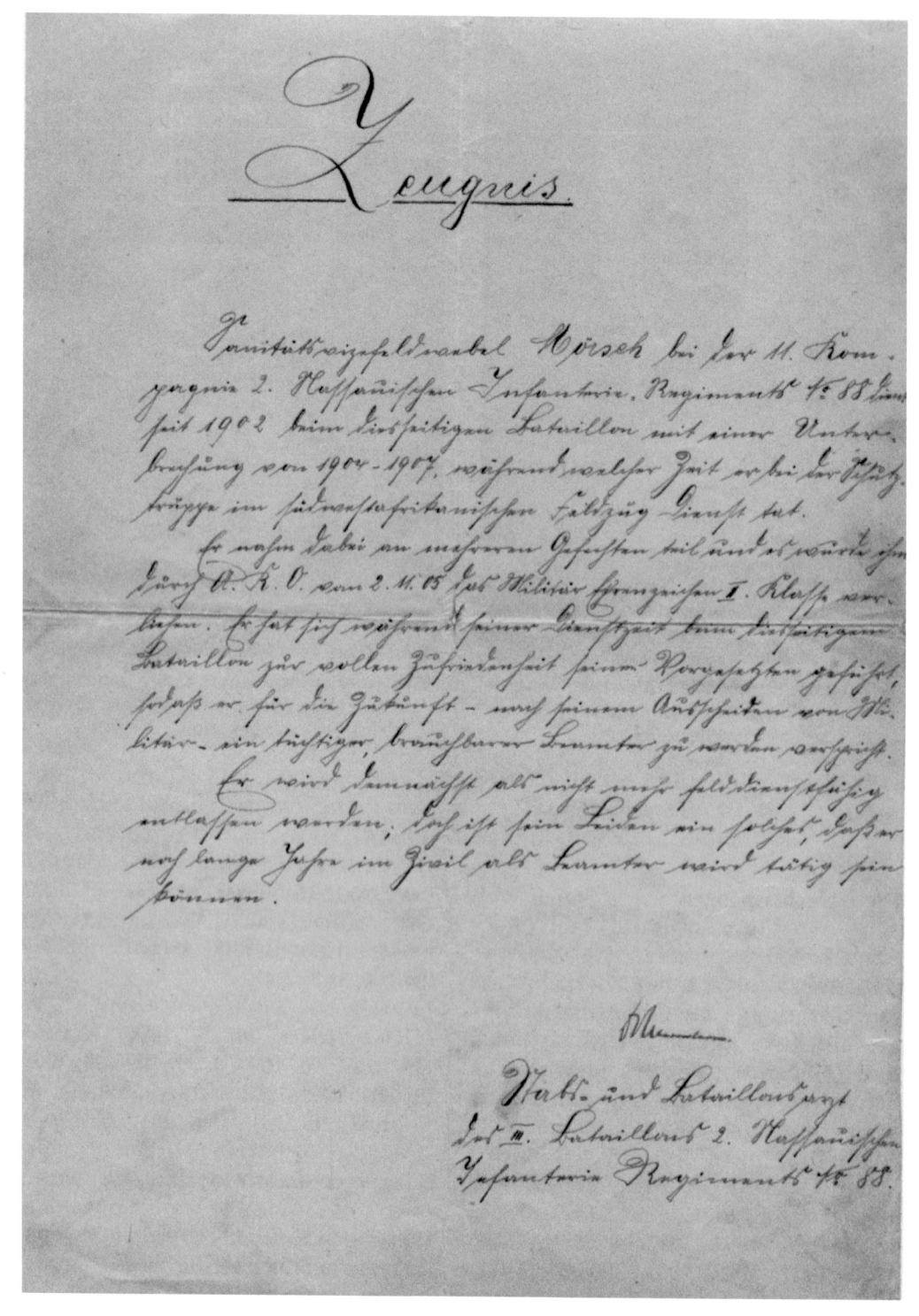

Testimony of Good Conduct for the Military Honor Award (*Militär Ehrenzeichen*), 2nd Class. *Sergeant* Moesch was with the German South-West African *Schutztruppe* from 1904 to 1907, and took part in various skirmishes.

Certificate from the Colonial Veterans Association awarding the Lion Order (Colonial Award) in Silver to *Hauptmann* Hartmann of the former German South-West African *Schutztruppe*, for service to the German colonies.

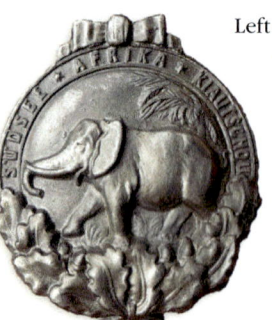

254  *Imperial German Colonial Troops & Police in Africa*

Unofficial Commemorative Medal for service in German South-West Africa. This Medal was produced in two different sizes.

Left: the Elephant Order *(Elefantenorden)* or *Kolonial-Abzeichen* was introduced by the Minister for Reconstruction in 1921 for those who served in the Colonies during the First World War.

Right: the Lion Order *(Löwenorden*, or *Kolonialauszeichnung)* was introduced by the German Colonial Veteran's League *(Deutschen Kolonialkriegerbund)* in 1922.

Photos: Helmut Weitze.

Two Veteran's Association Banners.

Banner photos: Jan Kube.

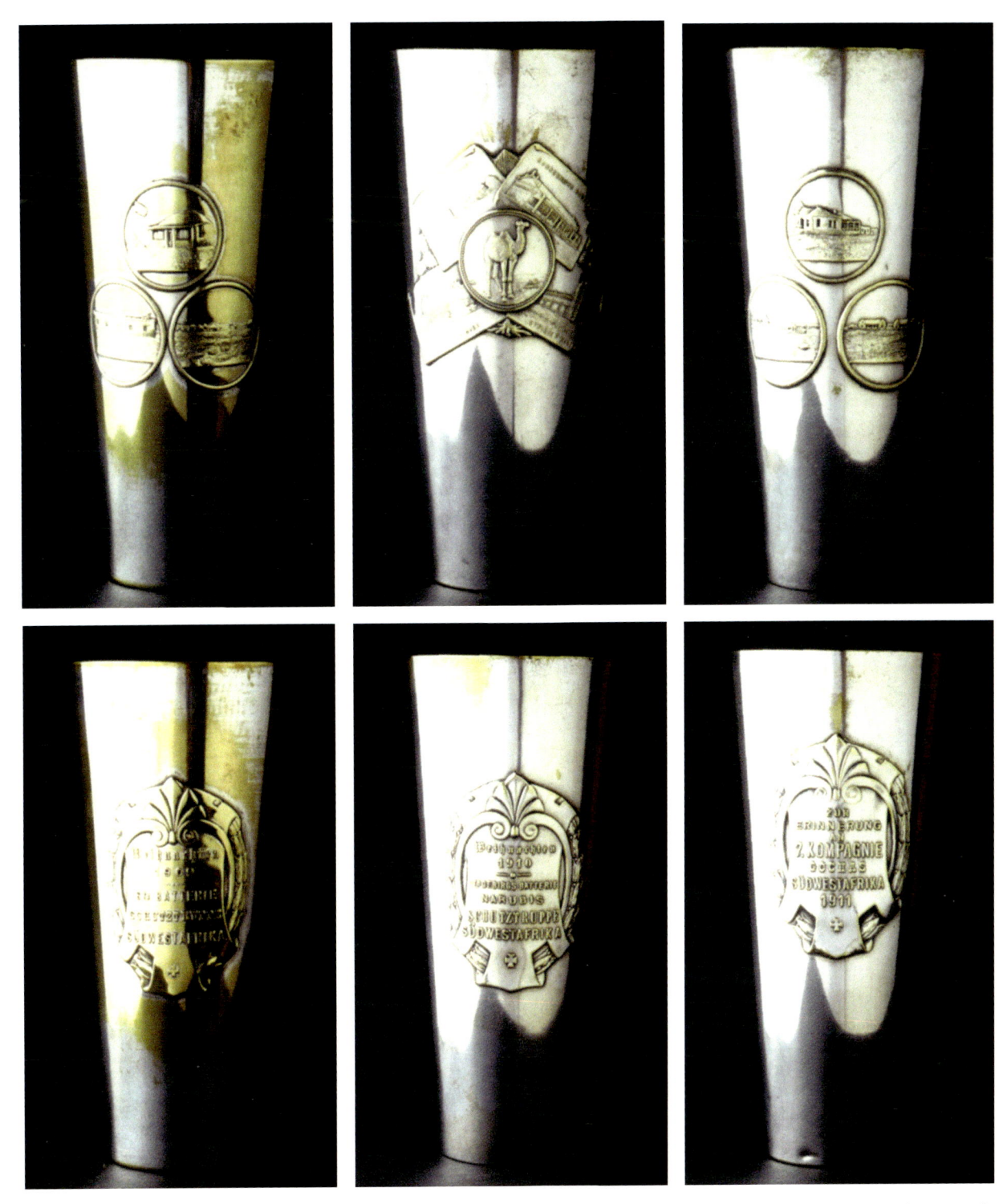

## Medals, Campaign Clasps, and Recognitions of Service

Memorial Cup for member of 6th Artillery Battery, German South-West African *Schutztruppe*, Christmas 1909.

Memorial Cup for 1st Mountain Artillery Battery member, German South-West African *Schutztruppe*, Christmas 1910.

Memorial Cup for 7th (Camel-mounted) Company member, German South-West African *Schutztruppe*, Christmas 1911.

General von Lettow-Vorbeck and *Askari* Soldiers in a Poster by artist Fritz Grotemeyer.
This Organisation supported *Schutztruppe*, *Polizeitruppe*, and *Marine* Veterans
as well as the families of those who died in service.

*Medals, Campaign Clasps, and Recognitions of Service*     257

Commemorative Plaque by sculptor Karl Mobius, depicting an African Soldier in Cameroon. Dedicated to "our beloved Zupitza on his 61st birthday, 1929". Doctor Maximilian Zupitza was a Senior Medical Officer who served in German East Africa during the Maji-Maji Rebellion, then in Cameroon, and was finally captured by British forces in Togo in 1914, along with the author's wife's grandfather, Doctor and *Leutnant der Reserve* Max Roscher. 15 x 10 cm.

Statuette of an *Askari* given to Dr. Heinrich Schnee,
former Governor of German East Africa.

Statuette of an *Askari*, gifted to a departing NCO of the
*Schutztruppe* for German East Africa.

Statuette of a German South-West African *Schutztruppe* Rider with Rifle at the ready.

# Medals, Campaign Clasps, and Recognitions of Service

A Commemorative Statuette for Dr. Schroedter of the South-West African *Schutztruppe*.

Statuette of a German South-West African *Schutztruppe* Camel Rider.

Statuette of an African Hausa Warrior
on horseback, a retirement gift to *Major* Künzlen
from his fellow Officers in the Cameroon *Schutztruppe*.

Bronze *Askari* Statutette, a gift from the
Working Group for Colonial and Maritime Interests in Leipzig.

Safe Conduct Pass from Kaiser Wilhelm II for Governor Schnee,
bound for German East Africa to take up his duties there. 17 June 1912.

**Reichs-Kolonialamt.**
**Kommando der Schutztruppen.**

Nr. M. 2837/12. K.B.
19790.

In der Antwort ist das Datum
und die vollständige Journal-Nr. dieses
Schreibens anzugeben.

Berlin W.8, den 31. Mai 1912.
Mauerstraße 45/46.

In Verfolg des Telegramms vom 31.Mai.

Unter dem erneuten Ausdruck aufrichtigster Teilnahme erfüllt das Kommando die traurige Pflicht, Ihnen auch noch auf diesem Wege die betrübende Mitteilung zu machen, daß nach einem Telegramm der Schutztruppe für Deutsch-Ostafrika Ihr Sohn, der Sergeant August Könnecke am 29.Mai abends, wahrscheinlich in Daressalam, an Malaria verstorben ist.

Der genannte Ort ist auf der beifolgenden Karte besonders kenntlich gemacht.

Über den Verlauf der Krankheit sowie die näheren Umstände, unter denen der Tod erfolgt ist, wird die Schutztruppe seinerzeit berichten. Sobald der Bericht hier vorliegt, wird Ihnen weitere Mitteilung zugehen.

Bezüglich Regelung des Nachlasses wird von dem Kaiserlichen Gouvernement in Daressalam das Weitere veranlaßt werden.

Ein von Seiner Majestät dem Kaiser entworfenes Gedenk-

An
Herrn Heinrich Könnecke, Landwirt,

<u>H e i n i n g e n.</u>

Letter from the *Schutztruppe* Command in Berlin, regarding the death of *Sergeant* August Könnecke in German East Africa, announcing receipt of his Memorial Page.

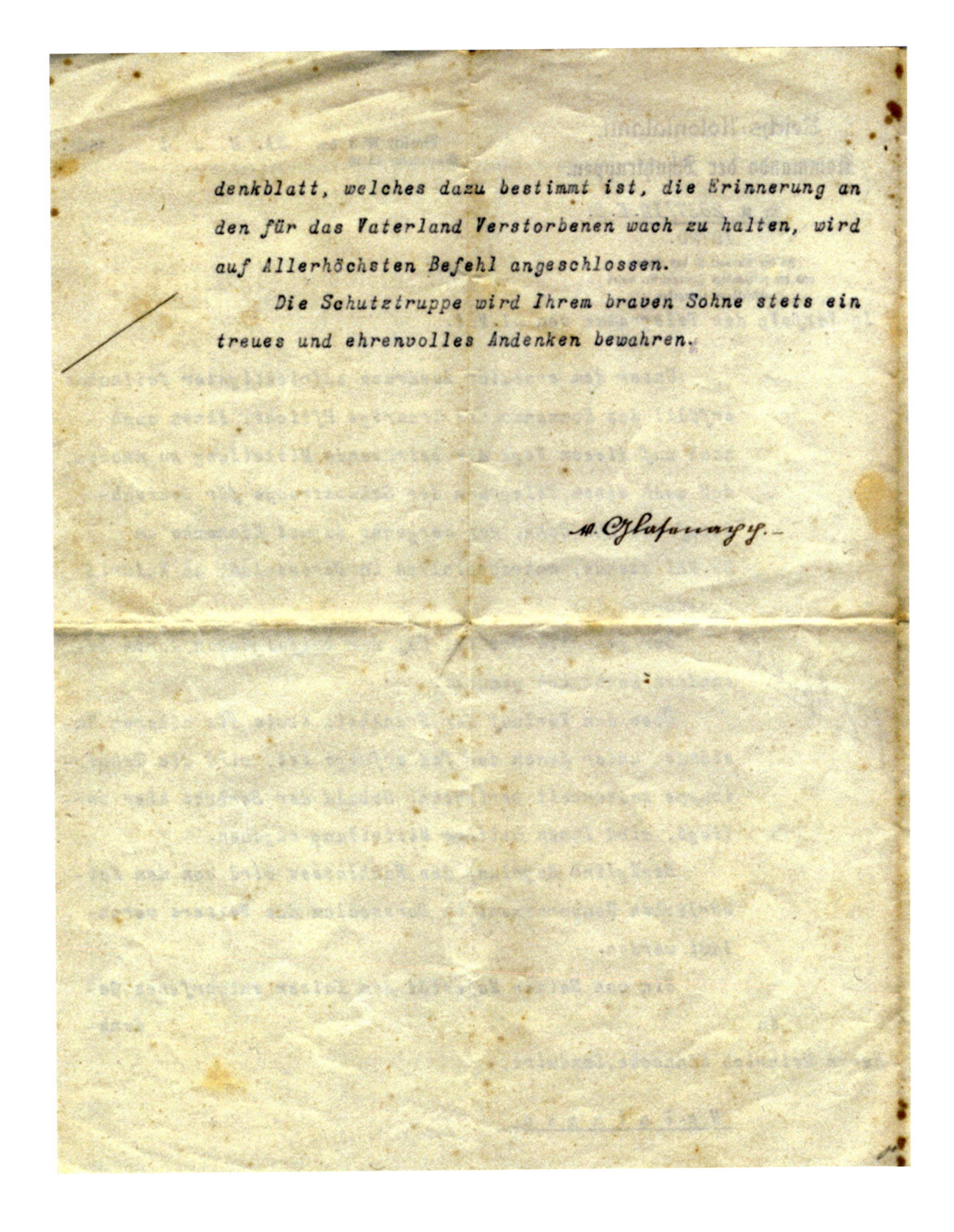

*denkblatt, welches dazu bestimmt ist, die Erinnerung an den für das Vaterland Verstorbenen wach zu halten, wird auf Allerhöchsten Befehl angeschlossen.*

*Die Schutztruppe wird Ihrem braven Sohne stets ein treues und ehrenvolles Andenken bewahren.*

**Gedenkblatt für die Gefallenen in Deutsch-Südwestafrika.**
Entworfen vom Kaiser.

Memorial Page for members of the *Kaiserlichen Marine* and *Schutztruppe*.

*Medals, Campaign Clasps, and Recognitions of Service*

Photo of a *Schutztruppe* Veteran in an elaborate commemorative frame.

**Vivat**
den
tapferen Verteidigern
unserer
afrikanischen
Kolonien.

Deutsch-Süd-Westafrika:
25.9.1914 Sandfontein.
Kamerun:
6.9.1914 Nsanahang.
1915-16 Verteidigung der Bergseste
Mora.

Deutsch-Ostafrika:
3.-5.11.1914 Tanga.
18.-19.1.1915 Jassini.
9.-11.5.1916 Kondoa-Irangi.
Okt.-Nov. 1916 Ugominji.

DEN
TAPFEREN
STREITERN
FÜR
DEUTSCHLANDS
EHRE

Zum Besten des Roten Kreuzes.

Photo: Helmut Weitze.

Gez. von Karl Werner, ehem. Schutztruppler.

Left: *Vivatband* commemorating African Campaigns during the First World War. Sales of

such ribbons benefitted the German Red Cross.  Above: Commemorative Print for Veterans of the German South-West African *Schutztruppe*.

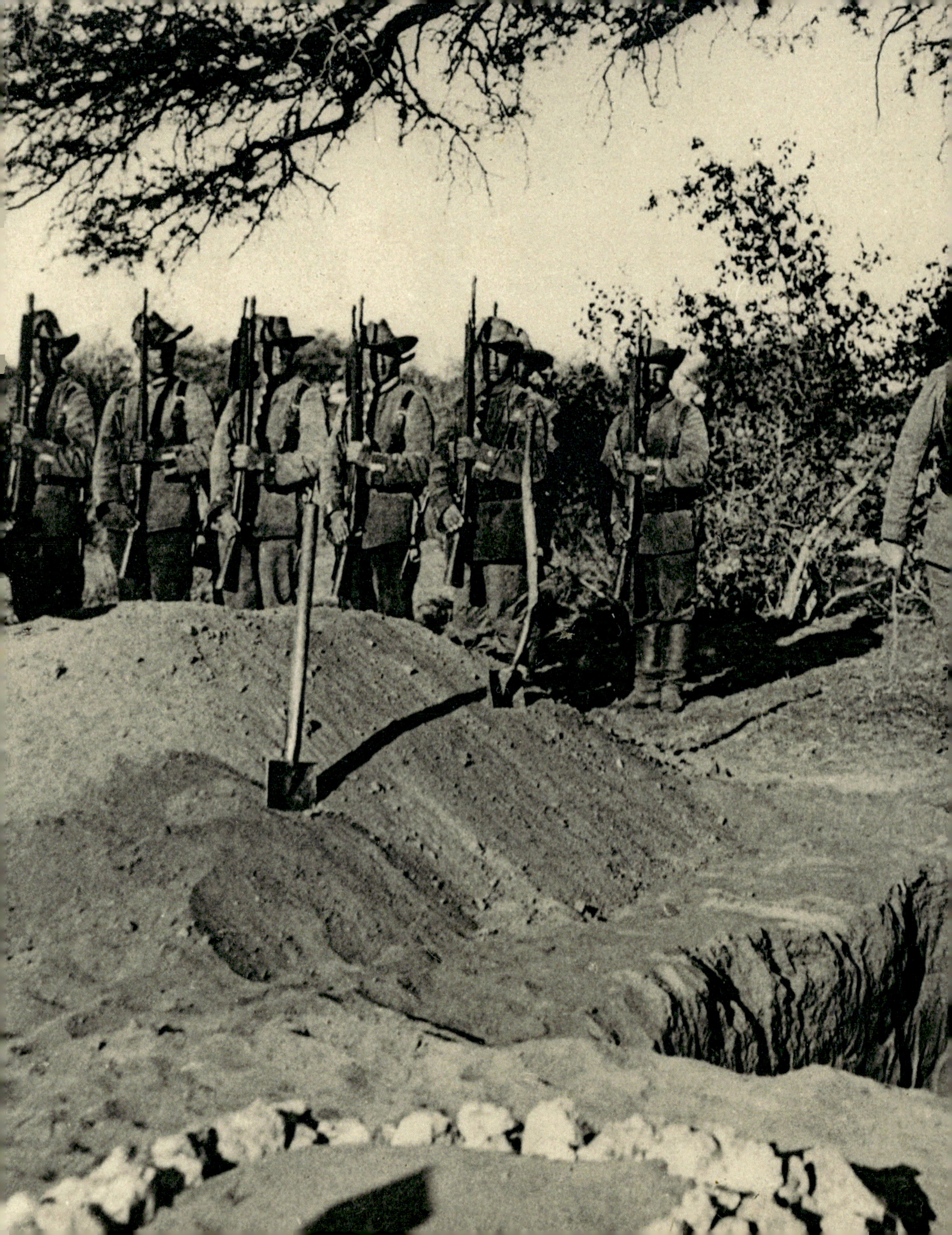

*Schutztruppe* Funeral for a fallen Comrade, German South-West Africa.

Above:
Imperial German Colonial Flag. 24 1/2" x 42" (62 cm x 106 cm)

Below:
P. Langhans' German Colonial Wall Map Nr. 1, showing Germany's Protectorates in Africa, ca. 1910.
62 1/2" x 83 1/2" (159 cm x 212 cm). Map Nr. 2 in this series shows Germany's Protectorates in the South Seas.

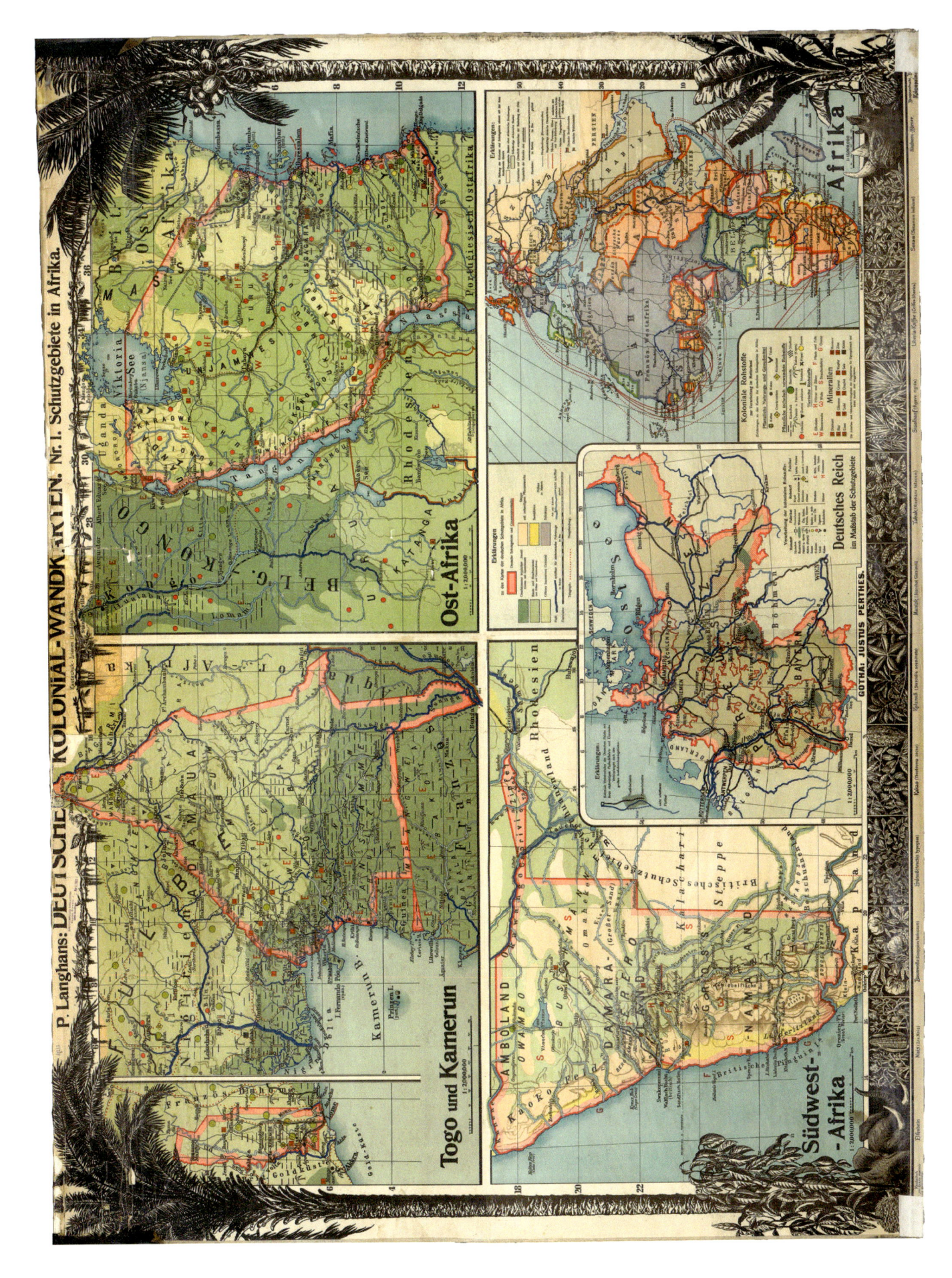

P. Langhans: **DEUTSCHE KOLONIAL-WANDKARTEN.** Nr. 1. Schutzgebiete in Afrika.

Ost-Afrika

Togo und Kamerun

Südwest-Afrika

Afrika

Deutsches Reich im Maßstab der Schutzgebiete

GOTHA: JUSTUS PERTHES.

*"Don't forget our Colonies"*

# Glossary of Schutztruppe Ranks

with First World War U. S. Army Equivalents

| | | | |
|---|---|---|---|
| *Gefreiter* | Lance Corporal | *Oberleutnant* | First Lieutenant |
| *Unteroffizier* | Corporal | *Hauptmann* | Captain |
| *Sergeant* | Sergeant | *Major* | Major |
| *Vize-Feldwebel* | First Sergeant | *Oberstleutnant* | Lieutenant Colonel |
| *Feldwebel* | Sergeant Major | *Oberst* | Colonel |
| *Leutnant* | Second Lieutenant | | |

# Selected Bibliography

Aming, Wilhelm :
- *Vier Jahre Krieg in Deutsch-Ostafrika*
  Hannover, Gebr. Jänicke

an anonymous *Schutztruppe* officer :
- *Meine Kriegserlebnisse in Deutsch-Südwestafrika*
  Minden, Wilhelm Köhler | 1907

Beckmann, Walther :
- *Unsere Kolonien und Schutztruppen*
  Berlin, Kyffhäuser Verlag | 1934
- *Die deutschen Schutztruppen in Afrika in ihrer gegenwärtigen Uniformierung*
  Leipzig | 1899

Frömming, Hans :
- *Die Bewaffnung der Kaiserlichen Schutztruppe von Südwest-Afrika*
  DWJ 8/1966

Hettler, Eberhard :
- *Organisation und Uniformierung der Wissmanntruppe*
  Mitteilungen einer Arbeitsgemeinschaft | 1932

- *Kolonie und Heimat :*
  Independent colonial weekly publication

Nigmann, Ernst :
- *Die Geschichte der Kaiserlichen Schutztruppe für Deutsch-Ostafrika*
- Photos, Maps, Documents compiled in photo albums by Hauptmann Nigmann.

- *Orden-Militaria Magazine :*
  68, 14. Jahrgang Oktober 1995

Laasch, Leue and Leutwein :
- *Mit der Schutztruppe durch Deutsch-Afrika*
  Minden, Wilhelm Köhler | 1905

Pohlmann, H. :
- *Die Kaiserliche Schutztruppe für Deutsch Südwestafrika*
  Feldgrau | 1955

Wagner, Rudolf (and Buchmann, Dr. E.) :
- *Wir Schutztruppler, die deutsche Wehrmacht Übersee geschildert*
  Berlin, Verlagsanstalt Buntdruck | 1913

Wernitz, Emil:
- *Vom Niemandsland zum Ordnungsstaat*
  Verband der Polizeibeamten für die deutschen Kolonien e. V. : Berlin | 1930

Wissmann, Hermann von :
- *Afrika, Schilderung und Ratschläge zur Vorbereitung für den Aufenthalt und Dienst in den deutschen Schutzgebieten*
  Berlin, Mittler und Sohn | 1903

Zache, Hans :
- *Das deutsche Kolonialbuch*
  Leipzig, Th. Rudolph | 1925

Photos, Maps, Documents from the archives of both Reinhard Schneider and Jeff Nelson

*May the spirit of 1914 remain with us, hopefully throughout long periods of peace!*
– Paul von Hindenburg

## About the Author

**Reinhard Schneider** was born in West Berlin in 1948 and has always remained true to his city. In 1969 his job as a steward on the passenger ship *"TS Bremen"* gave him a glimpse of the wider world. After that he was a sales representative in the beverage trade, and later worked for the City of Berlin. From 1977 to 1980 he served with the Volunteer Police Reserve, a paramilitary group trained to protect West Berlin.

His first marriage brought daughter Stephanie, born 1975, and grandson Moritz, born 2012. Marriage to his present wife Yvonne (Vonny) brought daughter Katharina, born 1981, and grandsons Theodor and Karl, born in 2010 and 2013. Together with his sons-in-law Glenn and Oliver, they form an extended family of which Reinhard is very proud.

For many years the author enjoyed painting *Zinnfiguren* (pewter figures) and building dioramas. His fascination with the colonies and tropical helmets has never waned, and to this day he relishes the hunt for a rare *Tropenhelm* to add to his collection. Reinhard's continued interest in these areas has led to his authoring two other German-language books:

- *Neuste Nachrichten aus unseren Kolonien in Afrika.* (**The Latest News from our African Colonies**), Carola Hartmann Miles-Verlag, 2010. 231 pages with 120 black and white photos. ISBN 978-3-937885-33-9

- *Die Tropenhelme der deutschen Wehrmacht – Afrika Korps.* (**Tropical Helmets of the German Forces in Africa**: *Heer, Kriegsmarine, Luftwaffe, Waffen SS, NSKK, Feldgendarmerie, DRK* – 52 pages, richly illustrated with more than 80 color photos of helmets and *Wappenschilder* and 34 historic black & white photos. Self-published, 2019.

# JAN K. KUBE e.K.
## KUNSTHANDEL – AUKTIONEN

Führender Spezialist
für

Historische Waffen

Helme

Uniformen

Orden und
Ehrenzeichen

Literatur

Historica

des 17.–20. Jahrhunderts

Jährlich mehrere große Spezial-Auktionen zu den genannten Sammelgebieten im Alten Schloss in Sugenheim mit internationaler Beteiligung.

**SCHÄTZUNGEN, EXPERTISEN, SAMMLUNGS-AUFLÖSUNGEN**
Ihr Experte auch im **BAYER. FERNSEHEN**, „KUNST & KREMPEL" seit 1991

**D-91484 Sugenheim · Altes Schloss**
Tel. (09165) 650 und 1386 – Fax 1292

Gegründet 1969 – Angebote immer erwünscht.

e-mail: info@kube-auktionen.de · www.kube-auktionen.de

KUBE seit 54 Jahren
1969–2023

# HELMUT WEITZE
## Militärische Antiquitäten KG
## Fine Military Antiques

As one of the leading dealers in Europe we are specialised in german Militaria from 1800 – 1945.

We offer medals & decorations, uniforms & insignia, hats & helmets, swords, daggers & bayonets, soldbooks & documents, wartoys from Lineol & Elastolin and much more. Please visit our homepage with over 20.000 articles.

**Weekly update every friday at 6:30 pm (german time).**

Helmut Weitze Militärische Antiquitäten KG, Neuer Wall 18, DE - 20354 Hamburg, Germany
Phone: 0049 40 / 471 132 0   Fax: 0049 40 / 353 563

www.weitze.net          info@weitze.net

Shop hours:
Mon. – Fri. 10.00 am – 6.30 pm
Sat. 10.00 am – 1.00 pm

*Hübsche Frau und ihre Schutztruppe.*